THE NEW ECOLOGY

THE NEW ECOLOGY

RETHINKING A SCIENCE FOR THE ANTHROPOCENE

OSWALD J. SCHMITZ

PRINCETON UNIVERSITY PRESS
PRINCETON AND OXFORD

Published by Princeton University Press,
41 William Street, Princeton, New Jersey 08540
In the United Kingdom: Princeton University Press,
6 Oxford Street, Woodstock, Oxfordshire OX20 1TR
press.princeton.edu

Jacket design by Amanda Weiss

ISBN 978-0-691-16056-6

British Library Cataloging-in-Publication Data is available

This book has been composed in Baskerville 10 Pro with
Futura Std Display

Printed on acid-free paper. ∞

Printed in the United States of America

10 9 8 7 6 5 4 3 2 1

CONTENTS

PREFACE

As a professor of ecology, I study nature and teach about what I have learned. Most of my teaching occurs in a university setting. But I also spend time teaching to enhance public awareness by speaking to various clubs and organizations. I find it highly rewarding. As a professor, I am also extremely fortunate because my vocation is my avocation. I am able to spend hours upon hours engaged in research, driven by my lifelong curiosity and wonder to discover the fascinating ways that nature works, and how humans with their various needs, wants, and values engage with it. I feel that as professors we have a tremendously important responsibility to teach current and future generations about these new findings; to excite them about ecology; to inspire them to take the torch and enrich the means and know-how that helps society achieve sustainable livelihoods. Most of us do this out of a personal ethic, a sense of obligation to uphold an implicit yet important social contract to provide society returns on its investment. As scientists, we strive to repay society's trust by doing our utmost to advance the kind of knowledge and practice that ultimately helps to solve environmental problems for the betterment of nature as well as for humankind.

In the course of doing all this, I continually find that ecology has come to mean different things to different people across society; and their views of how ecology can help society differ widely as well. This book is my way, as a professional ecologist, to explain to a broad readership what modern ecology is about, how it speaks to the many different viewpoints about science and nature that are held in society, and the new advances that ecological science is making in support of the health and welfare of all life on this planet.

Modern ecology—the New Ecology that I describe in this book—is the result of ecologists' expanding the frontiers of the life sciences with meaningful implications for society. Ecologists can now offer ways to reconcile, and perhaps even harmonize, different ethical views that humans hold toward nature, together with a scientific one rooted in ecological and evolutionary biology. Ecologists have come to embrace a new world view that nature should be viewed as the emergence of a perpetual ecological-evolutionary process of creation. This process provides the resilience needed to maintain sustainability in an ever-changing world. The deepest revelation of this comes from research focused on that creative process as opposed to merely marveling at and describing the magnificence of its products (e.g., the diversity of extant species). This new world view has implications for the ethical standing that we should extend to nonhuman life, especially as regards human management and conservation of the Earth's biota.

The goal of the book, first and foremost, is to show that ecology is a science devoted to studying the mysteries of the natural world. Ecological science endeavors to explain why different species exist in different geographic locations across the globe, why some are highly abundant and others not, and how environmental conditions and the nature of their interactions with each other control the functioning of ecosystems.

To do this, ecological science probes into the intricacies of how the genetic make-up and physiology of individual organisms determines their ability to cope with changes in the chemical, physical and climatic conditions of their environment. It relates how those intricacies shape the behavioral trials and tribulations of organisms as they vie for their share of limiting resources, or avoid becoming resources for other organisms, as they strive to make their way and grow and reproduce. It reveals how organisms fit together in communities through consumptive, competitive and cooperative relationships with other organisms. It traces how those interdependent relationships influence the flow of materials and energy, the production of food, the recycling of nutrients—all of which determine how ecosystems function. It tracks the ebb and flow of organisms and materials and energy across broad landscapes. It is a science that stirs awe as it demystifies nature's complexity.

As recently as fifty years ago, one could safely devote a career undertaking this kind of ecological study in vast tracts of wilderness largely devoid of humans or their impacts. It led to an ecological world view that humans and nature, for all intents and purposes, existed apart: as a human/nature divide, if you will. But the human population and its global reach now looms so large that it is no longer tenable to think of nature being unfettered by human influence. While wilderness does still exist, it is fragmentary owing to rising human domination of the Earth's space. The attendant rise in consumption of natural resources accompanied by and even driven by rapid technological development has done much to change the Earth. There is now cause for concern that the capacities of the Earth's ecosystems to sustain their functioning is becoming increasingly strained.

The New Ecology that I describe in this book is one that also aims to overcome the human/nature divide and tackle the issue of sustaining ecological functioning in the face of a growing human domination of the Earth, in a new epoch that has accordingly been called the Anthropocene. Ecological science is assuming a leading role to help guide the growing human enterprise, to support the needs and wants of a burgeoning human population, without jeopardizing nature's ecological functions that support humanity's needs and wants. It is doing this by advancing new concepts and theories to better explain the workings of an ever-changing world. This is what any science must do to remain vibrant and relevant. It means re-imagining how humans and nature come together to form functional systems, called socio-ecological systems. It means becoming more interdisciplinary in scope, to find creative ways to integrate the study of humanity (economics, anthropology, political and social science, religion and ethics) with the study of nature. It means ensuring that socio-ecological systems are sustainable. This essentially boils down to ensuring that they have the enduring capacity to be productive in the face of exploitation by making sure that species can exist and perform their functions over the long run.

This book sketches a portrait of a scientific field that is developing new ideas and ways to help humankind thoughtfully engage with nature in the interest of promoting a sustainable world. It addresses how human values and choices influence what ecosystems look like and how they function. It reveals how species, collectively through their functions, provide numerous services that humans have come to rely on to sustain their health and livelihoods. It engages in the ethical conundrums humans face when weighing the costs and benefits of various ways to exploit nature in support of their livelihoods. It reveals how choices to exploit nature in one part of the world can reverberate to the other side of the world. It also discusses how humans can use scientific principles to build new "natures" of their own making to transition to sustainable urban and industrial systems. It ultimately aims to show how humans, as part of socio-ecological systems, can use scientific principles to become thoughtful stewards in the interest of promoting sustainable livelihoods.

My hope is that the reader will come away with a fundamental understanding and fascination about the mysteries of the natural world. But more than this, I hope to offer inspiration and appreciation to think more deeply about how each of us engages with the natural world in our day-to-day lives, and consider the bigger picture implications of our values and choices.

While some historical context will be necessary to set the stage for certain ideas, this book will be largely forward looking to convey how ecology is growing as a science that can anticipate and help solve global environmental problems. The New Ecology offers the scientific means and know-how for humans to become better stewards of our environment, for the sake of humankind, as well as for the sake of the millions upon millions of species with which we share the planet.

This book would not have come together as it has without the help of friends and colleagues. Foremost, I wish to thank my editor Alison Kalett for her encouragement and guidance, to help me find my "non-professor" voice. I wish to thank Shahid Naeem, Reid Lifset, Mary Evelyn Tucker, Adam Rosenblatt, Lauren Smith, Rob Buchkowski, Julia Marton-Lefvre, Anne Trainor, and especially my wife Leslea Schmitz for offering discussion and feedback on various chapters, to ensure that they are technically accurate and that they are conveyed in a way that can be appreciated by a nonspecialist reader.

Finally, I thank my family and friends who have a genuine interest in sustaining nature's economy but are at a loss to demystify its complexity or to know what they can do as part of their lives to leave the world a more sustainable place. They have persisted in encouraging me to write a book that explains what I do in my profession and how it can help them become more thoughtful about their engagement with their natural world.

THE NEW ECOLOGY

CHAPTER 1

THE CHALLENGE OF SUSTAINABILITY

The word *north* has always allured me. Something about it connotes a distant, wild land with an arresting beauty that has persisted for time immemorial. Whenever I think of such a place, the Bristol Bay region of Alaska, with its breathtaking vistas of snow-capped mountains, crystal clear waters and lush, towering evergreens, immediately comes to mind. It is where large predators like bears, wolves, and wolverines still roam freely across the vast landscape, and where one can frequently see moose and caribou, bald eagles, and innumerable species of waterfowl. It is touted for having the greatest salmon run on the planet. Stocks of five species of salmon, that are among the last unthreatened stocks worldwide, use the region's headwaters as their nurseries. Each year, upwards of 40 million salmon set the rivers here ablaze in red as they undertake a spectacular migration back from the ocean to spawn in the area's headwaters. Along the way, the migrating salmon sustain ocean-dwelling killer whales, seals, and sea lions; and once in the rivers the dead and dying salmon provide key nutrients that sustain the many plant and animal species that make up the ecosystems within the region's watersheds.

The Bristol Bay region is also known for its geological formations that hold a mother lode of gold and copper, and a highly heat resistant metal—molybdenum—that strengthens alloys of stainless steel. The deposits of these metals, which lie directly beneath the very headwater streams used by the salmon, are so enormous that if mined they could double the inventory of the United States' copper and gold; and it would mean that the United States holds the world's largest supply of molybdenum. These metals sustain the high-tech manufacturing sector of our global economy. Gold is a key element in modern electronics including computers and cell phones. Copper is used for conducting electricity in power-grid distribution systems, residential wiring and electronics, and in motors that run all sorts of machinery. Molybdenum is an irreplaceable component of stainless steel used in surgical and medical equipment, and chemical and pharmaceutical manufacturing.

The desire to mine this reserve has led to much anxiety and acrimony. The debate centers on the wisdom of exploiting such an iconic and mystical place. Arguments on the one side hold that using the mineral wealth could boost the technological economy, including the innovative products and jobs that come with it. Counterarguments express worry that the mining activity would rapidly transform this wilderness area into a large industrial complex. This creates the risk that it will become a toxic wasteland that could drive the salmon to extinction, along with the species of birds and mammals that depend heavily on salmon for food. The issue is complicated by the fact that it cannot be resolved locally. Even if we never see this remote place firsthand, virtually anyone who clamors for the latest cell phone or computer technology or desires world-class health care would unwittingly have a hand in determining its fate by encouraging the exploitation of its minerals. This is little appreciated or understood because the ill effects of the mining will never directly harm most people. But the effects of transforming or destroying large wilderness areas can come back to influence humankind in a circuitous way. By virtue of supporting species and ecosystem functions, these wilderness areas also play a key role in regulating important Earth systems processes such as the global carbon cycle and thereby the climate.

This issue is emblematic of the kinds of tugs-of-war over nature that humankind increasingly faces across the globe. It is representative of the kinds of issues that ecological science is increasingly being called upon to help adjudicate. But it becomes complicated because of clashes between human values. There are those who have strong compulsions to subdue or tame nature's wildness and to exploit it, rationalizing that such a view benefits human economic health and well-being; there are others who revere its wildness for its pristine majesty and mystery, unspoiled by human presence. In either case, humankind typically does not view itself as being an intimate part of nature. Indeed, it has been difficult to imagine how humans could play a shared role in its inner workings alongside the other species that make up the natural world. To some, it would be uncivilized to do so. Wild animals and plants inhabit nature and so becoming a part of it would mean reverting to a seemingly primitive way of life. To others, it would be like intruding into a pristine and mystical place.

Either way, we have effectively created a human/nature divide. We are altering and controlling many natural areas expressly to suit our own purposes. In the interest of economics and commerce, we have transformed landscapes and ecosystems to enhance food supply, to extract ores and metals, to produce energy and building materials, and to reduce the danger from natural enemies like wild predators and disease. In the interest of conservation, we set aside some natural spaces as managed preserves and protected areas. But many of these spaces are mere fragments of their once vast size. Increasingly, there is less and less geography left on Earth that will not be influenced by one or another form of human agency. This can be reason for celebration or lament, depending on one's view of nature. Nonetheless, history has taught us time and again that humans will continue to exercise their impulse to transform and control nature. This was true thousands of years ago when societies transitioned to agrarian lifestyles, whereby they transformed wilderness into cropland and built infrastructure to irrigate those crops. It was true hundreds of years ago during the industrial revolution when societies expanded global trade and commerce, whereby they exploited wilderness to supply raw materials such as coal, iron ore and timber. It remains true today with the rise in urban growth and technological advancement.

So going forward, the looming question is: How can humanity engage with nature more thoughtfully and sustainably? From an ecological standpoint, sustainability means that ecosystems have the enduring capacity to be productive. This means ensuring that nutrients and water are replenished—recycled—at rates the meet the physiological needs of plants and animals to enable them to remain productive. It further means ensuring that species within ecosystems—the mind-boggling variety of microbes and plants and animals—can exist and fulfill their functional roles as interdependent members of food chains. Of course, any decision about how to do this must reconcile conflicting human values about nature. But the fate of these countless species and their interrelationships will inevitably hang in the balance. While ecological science may be called upon to provide a supporting role to help adjudicate the conflicts, it cannot tell people what values they should hold toward nature, nor what decisions they should make. The New Ecology can, however, encourage thoughtfulness by illuminating with scientific evidence how different decision options, based on those values, stand to influence the species that make-up ecosystems and their functioning. It thereby helps to ensure that any decision about sustaining nature is scientifically defensible.

My goal in this book is to show how modern ecology has grown to become a science in support of sustainability in the twenty-first century—an epoch known as the Anthropocene, in which humankind's actions will be the predominant forces shaping the world. This is not to say that, to keep pace, ecology has had to completely reinvent itself. I will show that it remains a science that remains true to its roots, fundamentally devoted to reveal nature's awe-inspiring mystery and beauty as it strives to understand the complexities of nature's inner workings. Indeed, an important discovery is how biological diversity—the variety and variability of life on Earth—is a central component of complexity that plays a key role in ecological functions that humans may draw upon to provide critical services in support of their livelihoods and well-being. But to remain relevant to the issues of today and tomorrow, it has become a science that also strives to re-imagine how human and nonhuman species can coexist and play a shared role in the workings of nature and the human built environment alike. As such, the New Ecology that I describe helps society overcome the human/nature divide by formulating scientific ways to integrate the study of humanity with the study of nature. I will discuss how these seemingly divided realms are in fact intertwined as socio-ecological systems—systems in which human political, cultural, religious, and economic institutions influence how nature works and how feedbacks from nature can instigate institutional change in a co-dependent way. This all means that we need to let go of traditional views that nature exists in some grand balance, and that humans have a persistent habit of disrupting that balance. Ecological science is revealing how nature is perpetually changeable, with or without human presence and, to borrow a turn of phrase from the environmental writer Emma Marris—it is sometimes rambunctiously so. I will explain how species have remarkable abilities to keep pace with change by continually evolving their physiological, morphological and behavioral capacities to cope. I discuss how preserving this evolutionary capacity is what is needed to keep ecosystems resilient, by which I mean that their functioning remains durable in the face of change. But the reader will also come to appreciate that human decisions that are made without any thought to nature's inner workings can stretch these evolutionary capacities beyond their limits. I will discuss how, in a modern world that is becoming ever more interconnected by global trade and commerce, the loss of evolutionary capacity in one location can have far-reaching consequences. That said, I also highlight how ecological principles can be newly applied to enhance the sustainability of human-built environments, such as cities and industries, in ways that can lessen societal demands and impacts on nature.

The New Ecology came to this place in a roundabout way. After it coalesced as a formal science in the early twentieth century, it advanced as two major subfields. One subfield, known as community ecology, grew from Victorian era natural philosophy that described the diversity and beauty of living beings, and a Darwinian evolutionary worldview that explained how those living beings came to be. It was fundamentally devoted to explain why different species existed in different geographic locations across the globe based on adaptations that were shaped by their competitors or predators. Community ecologists were also eager to know why some locations supported an incredible diversity and abundance of species and other locations did not. The other subfield—known as ecosystem ecology—grew from the earth sciences and was largely devoted to studying how materials and nutrients were cycled in nature. Ecosystem ecologists tirelessly accounted for the exchanges and storage of nutrients and materials among terrestrial, freshwater and marine reservoirs and the atmosphere. Even though they differed in focus, both subfields sought to do their studies in wild places devoid of human influence, because it was held that ecological and evolutionary processes were not anthropogenic in origin. Both subfields shared the world view that the biota and the nutrient cycles reached a grand balance. It was anthropogenic—humanly generated—affects that caused imbalances.

But in doing this, ecologists ironically perpetuated the human/nature divide we see today. This divided way of looking at the world progressed despite Aldo

Leopold's plea, during that time, that ecologists should not only study how nature works but should also apply their knowledge so that humans could see themselves as a part of nature's inner workings. Leopold was a professional ecologist, but he is best known as the father of modern environmental ethics. By integrating concepts from community ecology and from ecosystem ecology and relating to humanity, he was someone ahead of his time. He used his integrated view to articulate a basis for ethical engagement with nature in the interest of sustaining ecosystems and society. He also appreciated the conundrums ecologists always faced when reconciling the scientific study of nature with its conservation. The conundrum, which remains every bit as true today as it did then, is encapsulated in the following excerpt from his *Round River* (Oxford: Oxford University Press, 1953):

> One of the penalties of an ecological education is that one lives alone in a world of wounds . . . . . . An ecologist must either harden [his or her] shell and make believe that the consequences of science are none of [his or her] business, or [he or she] must be the doctor who sees the marks of death in a community that believes itself well and does not want to be told otherwise.

One could take the latter part of this quotation to literally mean that an important role for ecological science is to diagnose how bad humanly caused damages are and to highlight the dire consequences of further damage. This could be taken a step further to mean that ecological science should be marshaled in support of arguments that society desist in its over-exploitation and leave nature alone. And many ecologists have taken this position, speaking up to decry human destruction of nature. But doing this doesn't resonate with everyone. It may be highly appealing to those who behold nature with awe, simply because it exists in all its majesty and mystery. It can be less appealing to many others who may want to know what opportunities for human progress may be lost by protecting nature rather than exploiting it. Indeed, in some public circles, ecology has even become perceived to be a science in support of environmental activism against human progress, a science that perpetuates the human/nature divide.

The New Ecology I describe represents an effort to return to the kind of integrative worldview that Leopold had in mind. In fact, his use of a medical science and practice metaphor was meant to encourage the development of a parallel, integrated environmental science and practice. Such a science would, metaphorically, provide the means and capacity to diagnose how nature's ailments—the decrease in, or outright loss of, species diversity and ecological functionality—arise, of course. But in undertaking scientific studies to diagnose the problems, one also builds the kind of understanding needed to restore nature back to health; and more importantly help society minimize the risk of inflicting damages as it strives for greater technological and economic advancement.

But to relate more directly to modern concerns about human economic well-being and health, I would like to offer two ways of framing how ecology can deepen understanding about how humankind could play a role within nature, rather than apart from it, in the interest of building a sustainable world. One way is to take a kind of systems perspective that draws parallels with market economies. Here, nature is imagined as another kind of economy that is sustained by the production, consumption and transfer of materials by species. The other way takes an individualistic perspective by borrowing from new thinking in public health and medicine where sustaining the health of the environment is seen as going hand-in-hand with sustaining personal health and well-being.

* * *

Like a market economy that is built from a variety of sectors that provide specialized and essential services (e.g., agriculture, forestry, mining, manufacturing), nature's economy could be viewed as being built up from many kinds of species that together create another variety of sectors—called ecosystems—that also provide humans with essential services.

The variety of species that comprise different ecosystems is an important part of what is variously known as the diversity of life, biological diversity, or biodiversity. This diversity is what creates the opportunity for sustainability by ensuring that humans have ample clean and fresh water; deep and fertile soils; genetic variety to produce hardy crops; the means to pollinate those crops; and the capacity to mitigate impacts of gaseous emissions, among numerous other services. Together, the species and the bounty they produce could be viewed as a kind of capital: natural capital, nature's money in the bank. Spending that money faster than it can be accrued, or worse yet squandering it, sets things on a pathway to loss of essential services and eventual bankruptcy. This is another way of conveying what ecologists mean by loss of sustainability.

Envisioning sustainability this way requires the fundamental appreciation that we live on a finite planet. The space on the planet must be shared with other species if we wish to feasibly retain levels of the many ecosystem functions that are essential to providing services for society. Using this fundamental appreciation about nature's limitations, the New Ecology helps to envision how human dependency on ecosystem services connects to the functioning of those ecosystems. It can help to quantify and address trade-offs that inevitably arise because, in a finite world, expanded demand for any given service can lead to decreases in the ability of species within ecosystems to provide that or other services. Sustainability in the ecological sciences, as in other kinds of economics, is about finding appropriate trade-off balances, given limitations set by the scarcity of resources.

The management of biodiversity in nature's economy can also be metaphorically thought of as being akin to stewardship of portfolios of securities within mutual funds or similar instruments in market economies. At the very least, investors try to maintain a diverse portfolio of securities in order to mitigate risk and loss of financial return. But they also adjust their portfolios to maintain a desired level of return by keeping those securities that perform well in prevailing market conditions and letting go those that perform poorly. As prevailing conditions change, portfolios can be re-adjusted by selling off securities that perform poorly and buying back securities that are performing better. It is an effective way of adapting, by making sure performance and return stay on track as market conditions change. But there are limitations to extending this metaphor to natural economies because there are different consequences of letting securities go within a market economy and within nature's economy. In a market economy, a security that is excluded from a portfolio can be later included. That is, there is opportunity to adapt by making rapid reversals in investment choices by buying back those securities. In nature's economy, letting securities—species—go effectively means that those species may be difficult to bring back. This means losing the opportunity to make adaptive adjustments, which can lead to loss of return from vital functions and services as environmental conditions change.

Therefore, maintaining species diversity provides the capacity for societies to be resilient in the face of the challenges and changes that they must confront as they develop. Biodiversity provides this capacity because different species are adapted to live and function within widely different environmental conditions. This in turn creates the needed variety of choices and stewardship opportunities to sustain ecosystem functions and services.

For example, about forty different domesticated plants must be pollinated to produce the fruits, nuts and vegetables commonly found in the produce section of a typical grocery store. One could rely on a wide variety of natural pollinators to provide this service. This variety occurs because natural pollinator species have often evolved to specialize on one or a few flowering plants. However, in the interest of improving pollinator efficiency humans have instead developed a small industry around a single domesticated species—the European honeybee—that is capable of pollinating virtually all crop species. Beekeepers rear them in artificial colonies and transport them from one agricultural field to another to pollinate different crops throughout the growing season. However, domesticated honeybee colonies are now suffering catastrophic declines in their numbers. A key economic sector that humans rely on heavily is now jeopardized with this loss of pollinator service.

We could try to find substitutes such as the many native, wild pollinator species that collectively are equally capable of pollinating many of those crops. However, many of the plant species that make up the habitats of these pollinator species are often lost when every last bit of land is converted into agriculture to maximize the production capacity of the entire land base. Enlisting the services of wild pollinators to sustain agricultural production requires first re-balancing the trade-off between habitat conversion to promote crop production against habitat conservation to promote pollinator diversity. But unlike buying back a stock or bond where the transaction can be completed within hours to days, rebuilding the natural portfolio of pollinator species by restoring the plants species that comprise their habitats can take years to decades.

This is but one example of how awareness for and interest in maintaining a diverse portfolio of species in the first place could prevent the loss in natural adaptive capacity needed to avert changes in the fortunes in humanly dominated economic sectors (see chapters 2, 3, and 5). The point here is that the kind of human agency that leads to the pollination crisis should warn us of other kinds of pitfalls that arise as humans increasingly transform nature.

* * *

Achieving sustainability, I will argue, can also be imagined as coming from efforts to live productive and healthful lives. To get a sense of what I mean by this it is helpful to look to the health sciences and the way the field is trying to re-imagine the concept of human health and well-being.

Western medicine and public health has had a tradition of primarily treating patients when they have shown immediate symptoms of physical illnesses. As such, it is customary for medicine to apply its scientific know-how to diagnose the ailments and prescribe medicinal cures. There is, however, an arising effort in modern health science to consider human health in ways that extend beyond immediate illness, or ill-being. It involves identifying the conditions that promote healthful living. For example, Type II diabetes can be controlled by taking medication—insulin—to deal with its ill effects. But the risks of becoming diabetic in the first place could be reduced by adopting lifestyles aimed at well-being. This can include eating healthy diets, controlling weight gain, doing more physical activity, and reducing physical and mental stress. This new way of thinking, called positive health, emphasizes well-being by building on traditional areas of physical and mental health but expanding them to recognize that social, economic, and environmental conditions play an important role in determining how well we achieve physical and mental health. Public health and medicine have come to realize that focusing solely inward on the internal biochemistry and physiology of the human body dissociates humans from the environmental contexts that can make important differences in how well individuals are able to thrive and flourish.

Similarly, I would argue, achieving ecological sustainability demands healthful human engagement with nature. As such, it can be considered a scaling-up of the concept of positive health so that it applies to nature. It begins with concern about sustaining personal well-being. We certainly should not want to deliberately jeopardize our health by despoiling the natural environment with pollutants and toxic chemicals that cause respiratory ailments and cancer, or by degrading places in which we live through indifference or outright disregard for cleanliness and hygiene. The New Ecology is further showing that we can use species and natural places as part of preventative and curative medicine. For instance, promoting verdant urban forest can help offset air pollution and meet clean air standards and thereby reduce human respiratory ailments. It may therefore represent cost-effective alternatives to technological solutions designed to scrub pollutants out of the air. But sustainability entails more than taking actions to prevent such ill-being. It requires environmental awareness, a deliberate interest and ability to imagine how human choices and actions shape the urban and nonurban environment in which we live and how those choices may ultimately feed back to affect human well-being. It includes an awareness and concern for ecosystems in which we live and the diversity of species with which we share the planet. It calls for thoughtful stewardship over the environment in ways that can lead to healthful and resilient lives.

* * *

There is a growing sentiment that, as burgeoning human populations extend their use of the Earth's resources and open spaces to suit primarily their own needs, it will be impossible to keep all the species that occur everywhere today. So, society should consider triage approaches to help decide which species to keep and which species to let go. Given human nature, the inclination will be to pick the ones that suit our immediate needs or the ones we currently value the most. As I will explain (see chapters 2 and 3), such a view is inconsistent with thinking sustainably because it narrows the set of species that we rely on to provide a desired set of functions and services. And, as we have already learned, the New Ecology shows that such narrowing of choices means we will no longer have a diversified portfolio. Using this strategy risks losing adaptive capacity and future opportunity. The New Ecology has responded by taking Leopold's torch and helping to develop a scientifically informed ethic for a technologically advancing twenty-first-century society. The remaining chapters elaborate how the New Ecology is addressing these challenges.

CHAPTER 2

VALUING SPECIES AND ECOSYSTEMS

As someone who lives in a temperate part of the world, I tend to be equally fond of each season. But I have to admit that springtime can be a most welcome reprieve after a prolonged, cold, snowy winter. It is a period of reawakening in which animals and plants become impatient to get on with their business of living. As part of an annual ritual, spring means we will hear the vibrant trill of mating frogs, see the leaves of trees unfurl from their quiescent buds, and behold forest floors and fields brimming with color from a rich variety of blossoming wildflowers. The energetic pace of life becomes palpable.

The warming weather also triggers insect species to speed up their life cycles as they hatch from eggs, develop through several juvenile larval stages, and become adults that in turn lay eggs, the cycle thereby repeating itself. Some insect species do this as many times as the warm weather will allow; mosquitos are especially good at it. Females of many mosquito species pester humans in order to obtain blood meals that provide essential nourishment to produce eggs. Also, many, but not all, mosquito species carry diseases that infect humans while the mosquitos take their blood meals.

In the United States, several mosquito species belonging to the genus *Culex* have become especially notorious because they transmit West Nile virus disease to humans. The disease originated in Africa, but appeared in the New York City area in 1999. The disease quickly spread from this epicenter and by 2003 was found in all the states except Alaska and Hawaii.

West Nile virus infection goes unnoticed in most people. Others may come down with a mild fever and flu-like symptoms that pass after a few days. In the rarer worst cases, it causes swelling of the brain, called encephalitis, or swelling of the membranes around the brain and spinal cord, called meningitis, that can lead to permanent disorders or even death.

There is no vaccine or medicine to prevent the disease, making this an important public- health concern. And so, in another annual spring ritual we see public-health agencies ramp up very sophisticated scientific monitoring programs. These programs involve trapping and collecting adult mosquitos in the field to estimate their population sizes, followed by laboratory analyses that test for the incidence of the virus. When the virus is detected, it is reported to the media to inform the public to be vigilant and to take actions to avoid being bitten by mosquitos. When mosquito populations are large and the virus is highly prevalent, pesticides are sprayed into the environment to kill mosquitos and thereby control the potential for the disease to spread. Such control measures quickly help to allay public fear about becoming gravely ill, even though the risks of contracting West Nile virus disease to begin with are generally quite low.[1]

Mosquito control programs primarily target the larvae. Killing many of them before they can develop into adults lowers the chances that humans will be bitten. The larvae live in water. Thus, pesticides are applied to almost any water body in which mosquitos can breed, including ponds, lakes, woodland pools, wetlands, and

1 The U.S. Center for Disease Control (CDC) encourages pesticide spraying in densely populated metropolitan areas or areas with an abundance of mosquito breeding habitats. But is spraying warranted? CDC data indicate that during the five-year period between 2009 and 2013, West Nile infections reported within the entire United States have ranged from 720 to 5,674 cases. Of those cases, about half have led to encephalitis and meningitis. On average, 10 percent of those cases are fatal. Put another way, of all the reported cases, there is a 5 percent chance of dying from West Nile disease in the United States if a person becomes infected. This may seem a rather high mortality risk, but the percentage of victims is represented in terms those who are infected, not in terms of the population of the country as a whole. According to the 2013 U.S. Census Bureau data, the population of the United States was then 316,128,839. When considering even the highest level of infection reported between 2009 and 2013, the risk of anyone dying in the entire nation from West Nile virus infection in any single year becomes vanishingly small. By comparison, there is a 2,000 times greater chance of dying in the United States from some form of cancer each year; the chance of dying after slipping in the shower is 675 times greater. The chance of dying from West Nile virus infection in the United States is on par with the chance of getting struck and killed by lightning. I am not trying to be glib here. Disease outbreaks in human populations are rightfully causes for concern. But the benefits of pesticide applications to fight West Nile virus disease need to be weighed against potential hazards to other parts of the environment when the risk of contracting the disease itself is low.

estuaries. When such control measures are not enough, additional measures are taken to kill adults by also spraying pesticides aerially over large tracts of land.

Spraying pesticides into the environment is controversial. The worry is that the kinds of pesticides used and where they are applied can have side-effects that may impact many more species than just mosquitos, apart from the risks they pose to human health. Organophosphate pesticides like Malathion and Pyrethrin can be toxic to numerous other insect species, many of which, including the European honeybee, are highly beneficial if not essential to human economic well-being. Organophosphates applied to water bodies can kill invertebrates other than mosquitos and fish as well.

But pesticide applications must comply with strict legal guidelines that establish safe dosage levels based on scientific toxicity studies. And, indeed many of these chemicals are deliberately chosen because they quickly break down in the environment after they have been applied. Current scientific evidence indicates that the dosages applied in one-time control applications are safe. Humans have a higher risk of contracting West Nile disease than of becoming ill from a single pesticide application. Most animal species also are not directly harmed.[2] Nevertheless, concern about safety still

2 While studies find that animals often can tolerate very high concentrations, these findings are based on laboratory tests conducted on individual species. If individuals that are exposed to chemicals are subsequently submitted to conditions that resemble their natural environments, such as

abounds, especially given that pesticide residues can build up to harmful levels in the environment if they are repeatedly applied within a given year or spanning several years.

Inevitably, people begin to question what mosquitos are good for to begin with. Why not simply eliminate them altogether, and be done with these nuisance species and the attendant environmental hazards once and for all? Surely, given the countless numbers of other insect species living on Earth, there should be other species that could be suitable substitutes in nature's economy, without the harmful downsides. In fact, some pose the question: If mosquitos were beneficial to humans, wouldn't we have already found a way to exploit them?

This is the kind of sentiment that I started to address in chapter 1. It illustrates a very utilitarian, purely economic way of valuing nature, driven by a narrow focus on one small part of it. Such a narrow focus can disassociate humans from the larger ecological context in which they live. In the case example, their rush to deal with ill-being can cause humans to lose sight of the broader implications of their actions for their well-being. A major effort of the New Ecology has already

---

with their predators, a different picture emerges. For example, frog tadpoles exposed even to mild doses of the chemicals tend to be more sluggish and disoriented swimmers than nonexposed tadpoles. Such sublethal effects can cause exposed tadpoles to have a harder time evading their predators and thus they suffer higher predation than nonexposed individuals. The pesticide applications may be deemed safe because they do not directly kill animals, but the practice may predispose them to other risks of mortality.

broadened the scope of what is valued and how it is valued. Here I discuss some of the ecological scientific inroads that support a more complete valuation of nature.

* * *

The exact value that we place on anything, or whether we assign value at all, is personal. But the intrinsic, charismatic nature alone of large carnivores like bears and wolves and wolverines is enough for many to hold them in awe as the most magnificent beings on the Earth. No amount of money can even begin to compensate for the comfort or pleasure that we may take in knowing that these species exist somewhere on Earth. That is not all. Like all species on Earth, these species are the result of a chain of evolutionary events that can be traced back to the deep recesses of time in the Earth's history. The argument then goes: who are we to be so offhanded about what has taken eons to develop? Such an expression of value for the mere existence of species and their evolutionary heritage is steeped in the writings of Leopold and of John Muir before him.

Nowadays, however, many people increasingly desire—perhaps even demand—to express value in terms of the money they are willing to pay for the privilege or benefit of acquiring or using something that they want. Given that it is impossible to put a fair price on simply existing, does this mean that, in a twenty-first-century world, any one species could be worth nothing and

therefore expendable? For some, the answer clearly is a resounding NO! But for others, the answer isn't as clear-cut because it boils down to considering what other opportunities are available or are potentially missed by keeping the species. Given finite space on this planet, what is the opportunity cost—the lost chance to do something else—of maintaining vast open spaces in which magnificent creatures can roam widely and freely? In Bristol Bay, the opportunity cost is leaving multibillion dollars' worth of Alaska's minerals underground, valuable natural resources that could otherwise be exploited to support the desire for the latest technological inventions and advancement of society.

The rub here is that there is growing demand to base policy decisions on weighing only the financial costs and benefits of competing interests, on what economists call use values, rather than on non-use or existence values. But how does one reconcile a non-use value such as the beauty, uniqueness, and irreplaceability of a vast wilderness landscape like Bristol Bay against a use value based on the vast financial return from its exploitation? The answer, like it or not, is to try to assign a financial use value to species and ecosystems.

For instance, if people are willing to pay for the tourism opportunity to observe large predators or other species in their natural environments, they implicitly begin to assign a financial value to those species. The amount of money that people are willing to pay for this opportunity is a reflection of the value they place on these species. The more money that people are willing to pay to see them, the higher the value that is assigned to those species.

However, the financial return from tourism to see nature often pales in comparison to the financial return from exploiting nature for material gains. In Bristol Bay, for instance, net annual tourism revenues, including supporting employment, are calculated to run between $300 million and $500 million. By contrast, estimates suggest that if mining were permitted, the net annual revenue from the extracted metals alone could range between $2.9 billion and $3.3 billion. These sorts of considerations, however, reflect a narrow view of nature's value, because the valuation is merely based on material goods for which trade or sales in markets can set prices based on supply and demand.

Ecological science is now showing that more is at stake. Humans "exploit," or rather rely heavily on, nature in another way, often without realizing it because as a society we aren't required to directly pay anything for it. Because we aren't expected to pay anything, we take it for granted, even though our very existence relies on it. I am referring here to the service value that nature provides to humans, such as providing clean water, fresh air, and deep and fertile soils. Species within nature's economy provide this service value through their functions. Species are integral functional components of nature's economy. By producing materials, by consuming materials, by decomposing materials, they sustain nature's economy. These functions collectively ensure that nutrients are continually recycled and distributed throughout ecosystems to maintain overall ecosystem productivity. All the while, this functioning supports societies by providing services that support their existence. Thus, the value of extracting minerals, within a short time horizon of thirty to fifty years, must be measured against the value of nature's services provided by that same location, which, for all intents and purposes, is perpetual. The problem is that, when making decisions, the economic value of something that lasts into perpetuity is often discounted far more heavily than something that lasts for a shorter period of time. Insights from ecological science, however, are helping all of us to rethink the wisdom of such approaches to decision making.

* * *

At a primary level, all of us would surely starve if we didn't have species to build up and maintain the Earth's soils that we in turn use to grow food. These often microscopic species include bacteria and fungi, and small invertebrates like beetles, earthworms, and millipedes. They are without doubt the least charismatic of all species on Earth. Many would find it hard to take comfort or pleasure in knowing that they exist at all, except that, without them, waste materials would not be broken down; nutrients would not be replenished; and plants and animals would be unable to thrive. In short, ecosystems could not be sustainable. Human ingenuity is not even close to building the kind of technology that can substitute for these species, let alone reproducing the evolutionary conditions for these species in that habitat.

These species and their functional roles tend quite literally to be out-of-sight and out-of-mind because they live within the soil. At first glance, soil could be viewed as just a bunch of rocks and dirt. A sort of "black box" into which nature's waste, like plant leaves and stems, animal wastes, and dead animal carcasses, is deposited and somehow magically turned into elemental nutrients that fertilize new plant growth. But a closer look reveals a world teeming with life. A dizzying variety of tiny invertebrates and fungi and bacteria are found there, and they play an important role as decomposers. Working together, decomposer species like earthworms, beetles, millipedes, and woodlice initially shred animal and plant matter into smaller particles of organic matter—the soil humus. These species also improve soil quality by aerating it as they burrow, by depositing fecal pellets in it, and by improving habitat conditions for bacteria and fungi. Bacteria and fungi then take over. By instigating an array of chemical reactions that ecologists are only now beginning to fully understand, these species break down this organic matter into constituent molecules and further transform them into elemental forms that nourish plants. Bacteria can be exceedingly plentiful. A teaspoonful of highly fertile soil is estimated to contain between 100 million to 1 billion individuals. Fungi can symbiotically network their branching underground filaments, called hyphae, with plant roots. They thereby develop huge underground conduits for translocating nutrients from plant to plant across a wide space. In a world with finite resources, the processes of recycling and redistributing nutrients in the soil are what keep nature's economy sustainable and productive. Soil fertility and productivity, in turn, are key to humanity's ability to feed itself.

Soil is the foundation for building up the structure of ecosystems as well. Soil provides the substrate for plants to root themselves firmly in place. Plants in turn provide habitats and food for animals. This structure, called a food chain, is the scaffolding of nature's economy. In this chain, plant species (primary producers) draw up water and nutrients from the soil and draw down carbon dioxide from the atmosphere, and stimulated by sunlight undergo photosynthesis to build plant tissue. Herbivore species (primary consumers) in turn eat that plant tissue, manufacture and produce animal tissue, and are themselves eaten by predator species (or secondary consumers) that further produce animal tissue. Individuals of plant, herbivore, and predator species that are not eaten by their respective consumers eventually die from aging, and the chemical elements of their bodies are recycled by decomposer species in the soil. Accordingly, we can't look at soil, nor the species within food chains that directly or indirectly depend on it, as separate entities. All are parts of the natural economy, linked together by their feeding interdependencies. Hence, any action that humans take to change or affect one part of the economy reverberates through the rest of the interdependent chain.

Ecological science, now and more than ever in its history, teaches us that, when valuing and exploiting nature, we need to think about the implications for the entire economy, the ecosystem. Whenever we privilege certain species over others, we risk distorting functions and compromising sustainability. Large carnivores cannot be conserved without somehow providing them with ample herbivore prey. Herbivore prey species cannot be conserved without providing ample plant forage and habitat for them. Plants cannot be conserved in the absence of productive soils. And so forth. This could be understood as a restatement of classic "systems thinking" in ecosystem ecology, that the whole is greater than the parts. This classic view, however, did not consider how the parts—the species and their mechanisms of interacting—could predictably contribute to the whole. The "whole" was treated as having an unknowable, ethereal nature. The New Ecology is increasingly revealing the importance of knowing the intricate mechanisms driving the dynamic interplay among species and what it means for the entire system if they are disrupted.

For instance, the boreal forest region, a vast wilderness in northern Canada and northern Russia, covers

10 percent of the Earth's land area. It is typified by evergreen and deciduous trees species like spruce, pines and aspens. It is considered a safe haven for large carnivores like bears and wolves and wolverines. Boreal forests also store more carbon than any other terrestrial ecosystem on Earth, twice as much as in tropical forests. This wilderness area thus has extraordinary value in an era during which concerns abound that human activities are causing climate warming by releasing carbon-based greenhouse gasses to the atmosphere.

Boreal forest trees, like all plants, take carbon dioxide back from the atmosphere in order to produce stems and leaves and roots. But, unlike in the tropics, most of the boreal forest carbon is not stored within these living plant parts. Rather, it is stored in the soil as dead organic matter, once leaves and branches and roots are shed from the plants. The soil becomes a huge carbon repository because cooler conditions slow down the activity of soil organisms, which in turn slows the breakdown of all animal and plant matter. This process by which carbon becomes stored within the ecosystem is known as carbon sequestration.

Boreal plants are an integral part of the food chain as a food source for large herbivores like moose. If moose populations become very large, they can have a significant impact on the abundance of trees, which means that less carbon is taken out of the atmosphere, and less carbon is eventually sequestered in the ecosystem. But large carnivores like bears and wolves often keep moose populations under control. By reducing the impact of moose on boreal trees, these carnivores indirectly benefit the plants and thereby enhance the storage of carbon in boreal soils. Calculations have shown that, across the entire boreal region of Canada, protection of these large carnivores and their associated roles within nature's economy could enable boreal ecosystems to sequester enough atmospheric carbon dioxide each year to offset all of Canada's annual carbon dioxide emissions from burning fossil fuels. In the global carbon budget, this is not a trivial amount. Among the 190 or so countries in the world, Canada is ranked among the top fifteen largest emitters of carbon dioxide.

Species like bears and wolves are mostly valued for their charismatic nature, or perhaps reviled for killing game species, like moose, that humans may value more. But as a part of the food chain, wolves and bears also provide a huge environmental service to society. Ecologists have been able to provide evidence of this service by conducting large-scale scientific experiments in places like Isle Royale National Park in Michigan. This protected wilderness at the western end of Lake Superior, 535 square kilometers in extent, has been home to both moose and wolves for about seventy-five years, after a harsh winter allowed wolves to cross the ice from the mainland. During more than fifty years of continuous scientific study, it is the location of one of the longest-running analyses of how interactions between a predator and its prey influence ecosystem functioning.

Experiments there fenced certain areas to exclude moose for long periods of time. By comparing the fenced area with similar-sized areas where moose were free to roam, ecologists were able to measure how moose presence and absence caused changes to the ecosystem. Ecologists also measured how predation controlled moose abundances naturally over time. The study provided a portrait of how much wolves benefitted ecosystem functioning through their control over the moose population. Similar experimental studies show all of us that we cannot adequately value the importance of species by their existence alone. Their full value to society can be appreciated and quantified only relative to their absence. That is, valuation of the complete wealth that nature's economy offers requires conducting experiments that enable an accounting of the missed opportunity cost of losing the species—of accounting the cost in terms of lost environmental services—in addition to their existence value and their market value. There are other examples that provide valuable lessons about how the loss of a single species can have far-reaching ramifications for ecosystems—and for the global carbon budget.

As natural wonders go, perhaps the most awe-inspiring is 1.2 million wildebeest flowing across the vast Serengeti grassland during their annual migration. It would be a tragedy to lose them. But we almost lost them when, decimated by disease and poaching, their numbers crashed to 300,000 prior to the 1960s. Much of the Serengeti ecosystem consequently remained ungrazed. The accumulating dead and dried grass in turn became fuel for massive wildfires. The fires annually burned nearly 80 percent of the area, making the Serengeti a major regional source of carbon dioxide emissions to the atmosphere. Wildebeest population recovery, through disease eradication and anti-poaching enforcement, restored the grazing system and reversed the extent of the wildfires. Grazing now causes much of the carbon in grass to be released as animal dung that is in turn incorporated by insects into soil reservoirs that are not likely to burn. The Serengeti ecosystem has now reverted to being a net carbon dioxide sink. The amount of carbon stored annually is estimated to offset all of East Africa's current annual fossil fuel carbon emissions.

After being hunted to the brink of extinction during the fur trade, legal protection and management efforts have helped sea otter populations to grow and to rebuild coastal forests of kelp by regulating herbivorous sea urchins that devastated these forests. Recovering sea otters to historic levels, even though along only the thin stretch of coastline from Vancouver Island to the western edge of the Aleutian Islands, has created the potential for storing 6 to 10 percent of the annual carbon released from fossil fuel emissions in British Columbia, Canada.

More broadly, coastal ocean ecosystems offer some of the greatest opportunities for carbon storage. Salt marshes, seagrass meadows, mangroves, and continental shelves are some of the world's most important carbon sinks. Often regulated by predator-driven knock-on effects (the impact of carnivores on an ecosystem), these marine sinks can store carbon up to forty times faster than can tropical forests. But overfishing of predatory fish and crabs in places like Cape Cod saltmarshes has resulted in an explosion of herbivore crabs and snails, with large swaths of marsh dying off and the subsequent erosion of tide-exposed sediments. Along with this comes the loss of hundreds of years of stored carbon, and loss of the future possibility for sequestering carbon. On continental shelves, fish and invertebrates such as starfish and sea cucumbers, precipitate carbonate minerals in their intestines as a byproduct of physiological processes to prevent excess calcium assimilation into their bodies. Globally, this annual production and release of carbonate as waste to the ocean floor, where it is stored, is estimated to equal the annual fossil-fuel carbon emissions by such countries as Brazil, the United Kingdom, or Australia.

Ecology is learning more about how one or a few animal species can help determine the amount of carbon that is exchanged between regional ecosystems and the atmosphere. Adding up these regional contributions globally may create a portfolio of solutions that together can meaningfully help slow the build-up of atmospheric carbon and thereby to slow the pace of climate change.

But the fact of the matter is that large carnivores are never highly abundant within an ecosystem relative to the sheer numbers and total mass of herbivores, whose numbers in turn pale in comparison with the abundance of plants and microbes within an ecosystem. Because of the rarity of bears and wolves, these species, or other rare species, are often overlooked when accounting for the role of living creatures in nature. When species are overlooked—and thereby have their value implicitly downgraded—they become casualties of the human enterprise. The emerging lesson here is that the rarity of any species could belie its importance and value. Rare species can potentially instigate large effects within an ecosystem. These effects of carnivores are also called trophic cascades because they can trigger the circuitous feedback alluded to in chapter 1.

* * *

Like Bristol Bay, the boreal forest wilderness is also prized for its wealth of mineral resources, as well as for its vast timber supply, and its oil and gas deposits. It is inevitable, then, that conflict would arise over whether to maintain the region as a vast intact wilderness or to break it up by extracting its abundant natural resources. Resource exploitation is arguably critical because it drives the market economy that supports human well-being. The value of such exploitation is easily quantifiable because natural resources are material goods that can be priced based on market valuations. But, in an interesting twist, we have now learned that there is an opportunity cost—in the form of decreased carbon sequestration—associated with exploiting the boreal region for its natural resources instead of keeping it as vast, intact wilderness.

Natural resource exploitation disrupts species' habitats and thereby diminishes the ability of species, especially large carnivores like bears and wolves, to thrive. This may compromise the sustainability of the natural economy of the boreal ecosystem, especially the ability to offset carbon dioxide released from human activities that drive the market economy. For instance, once released from predation pressure, moose populations can increase and overbrowse growing trees. Estimates show that even a small rise in moose density from fewer than one to fewer than two animals per square kilometer (the lower end of recorded moose densities for boreal forests) is sufficient to reduce soil carbon storage per unit area by 10 to 25 percent.

Furthermore, fragmenting the boreal forest, by cutting down large swaths of trees and building extensive road networks, can lead to soil warming. This allows soil organisms to become more active, following which the soil organic matter becomes decomposed more rapidly. This causes the carbon stored in the soil within a given area to be depleted, releasing carbon dioxide to the atmosphere. If resource exploitation is conducted on a large enough scale, it could turn large parts of the boreal region into a source of atmospheric carbon, rather than being a carbon sink.

The value of vast, intact boreal wilderness, far above and beyond its scenic beauty, is that it offers the spatial extent needed to allow species to uphold their functional roles in nature's economy and thereby sustain critical life-support services. Wilderness is not forsaken hinterland. Humans are inextricably tied to it wherever we live on the planet. We are directly tied to it because we depend on natural resources. We are indirectly tied to it through the circuitous reverberations and feedbacks caused by that resource exploitation. These areas, while largely uninhabited by humans, are nonetheless key to offsetting industrial greenhouse gas emissions that result from human activities that are concentrated in more densely populated areas.

If society were willing to weigh policy decisions by accurately pricing carbon, then the environmental service value of mitigating atmospheric carbon through sequestration into wilderness areas can be placed on an equal footing with the value of exploiting natural resources from those same areas. Ecologists and economists advocate for carbon pricing, but not for the purpose of precluding resource exploitation or taxing the generation of wealth in the market economy. Rather, such pricing increases awareness of the value of nature, and it expands the ways that nature is valued, and of the opportunity cost of losing it.

* * *

The New Ecology is increasing scientific know-how in order to help adjudicate the various competing demands on nature. This knowledge is based on facts gathered during field research. Foremost, that knowledge helps us to appreciate that, often by default, society plays a zero-sum game when dividing up landscape space among competing uses. Whenever society chooses to dedicate land space solely for agriculture to feed the hungry or to produce biofuels, or to extract minerals for technological innovation, or to build resort hotels on seacoasts for tourism, its choice in fact gives up some amount of environmental service that that space had provided. The New Ecology shows that society needn't do so. For example, to adapt to the risk of losing agricultural productivity when domesticated European honeybees decline, it is possible to assign some agricultural land as habitat for wild pollinator species. This is one example whereby humans could live compatibly alongside biodiversity on the same land space and promote multiple uses like agricultural productivity, conservation of species and of their functional roles, and their associated environmental services. Focusing on biodiversity and its conservation is key to maintaining the sustainability of an interdependent nature and society.

CHAPTER 3

BIOLOGICAL DIVERSITY AND ECOSYSTEM FUNCTIONS

I sit in my car in the twilight of dawn before I pull up my waders, grab my fly-fishing rod, and enter the river to try my hand at catching some trout. Fly-fishing is a great way to put ecological scientific knowledge into practice. It requires an ability to understand the physical structure of a river: to read the flow of the water—distinguishing between the shallow fast-flowing riffles and the deeper pools—and thereby to locate the habitats where a prized fish might be lurking. It requires knowing the identity, the life cycle, and the habitat requirements of the many insect species that live in the water and on the adjoining shoreline, any of which are potential prey for the fish. It requires observing where the fish rise and which insect species they snap up at this particular time, so that I can select an artificial one from a fly box that best resembles the real thing. It requires emulating the flight and swimming behavior of that insect when casting the artificial fly in order to trick the fish into thinking it is indeed the real thing.

But I come here for more than just the opportunity to fish. As I step into the water I pause a moment and look up into the forest that borders the shoreline. I admire the kaleidoscope that the flecks of rising sunlight cause while penetrating the small gaps within the dense forest canopy. The rising sun is nature's alarm clock. The many species of songbirds that call the riverine habitat and adjoining forest their summer home awaken and begin their morning chorus. There is not a person in sight, anywhere. The only sound comes from the singing birds, the gurgling, crystal-clear, flowing water, and my breathing. I take a deep breath and take in the fresh smell of pine air. This stretch of the river is my place of solitude. Here I can decompress and forget the sometimes frenetic, stressful pace of urban life.

Some would say that this is all well and good if one has the time and money to travel to such a precious place of wilderness solitude, but that, for most people, it is a luxury that they can ill afford. Ironically, however, my place of "wilderness" solitude is only a fifteen-minute drive from my house within the heart of the city in which I live. It belongs to a larger urban greenspace, a corridor that runs for miles to connect several towns and cities. It is an urban watershed—a linked aquatic and terrestrial ecosystem—in which the river and the surrounding landscape through which it flows are an integrated whole. It would not be there but for the vision of urban planners in the city's early history. They understood the importance of protecting such a watercourse within the confines of a larger region that would be increasingly transformed by humans. Their early actions to sustain this ecosystem have paid-forward many times over to generations of urban dwellers. These kinds of benefits are known as ecosystem services.

Ecosystems provide many services that species perform by simply living and functioning within them. The functions in which species participate are soil formation, nutrient cycling, pollination, and production of plants and animals in food chains. There are provisioning services, which include the production of food, timber, fuel, and fresh water. There are regulating services, which help to moderate or stabilize climate, prevent flooding, purify water and air, and minimize or prevent outbreaks of disease. There are cultural services, which speak to recreational, aesthetic, and spiritual needs.

My place of solitude provides several of these services all at once. It includes services that attend to my personal needs, and it also provides services of value to the greater urban community in which I live. For me, it represents a way to think about my personal connection to nature in terms of positive health benefits. Furthermore, drawing connections between ecosystem functions and ecosystems services can make the concept of sustainability less nebulous. As an ecologist, I find that it offers tangible ways to translate science into practice by first being amazed by and appreciating the intricacies of nature, and then by revealing the many threads that tie humans to nature through such intricacies. Drawing such connections illustrates why it is important to ensure that ecosystem functions endure.

For instance, basic to my own needs, the river offers several cultural services. Fishing is an important recreational activity. I am able to do this in an aesthetically pleasing surrounding. This place can attend to spiritual needs as well. It is balm for the human spirit. The experience and enjoyment of fishing in a place of solitude gives my mind a break, allowing me to meditate and forget the day-to-day challenges and anxieties of my work. Such relaxation leads to physiological changes, not the least of which is a lowering of my blood pressure that can persist long after I have left. Solitude can also offer, for some, a feeling of being in a safe haven. I have read countless heartrending testimonials from reporters and victims who have witnessed horrendous world events. They describe how they were able to cope because of the refuge and solace afforded by places similar to my urban wilderness solitude. The connection to nature in such urban places, however personal, is how one begins to overcome the human/nature divide at an individual level. It offers a way to begin connecting each of us to environmental contexts that contribute toward achieving positive health outcomes.

Positive health outcomes are also realized at a broader societal level by nature's provisioning services. The New Ecology has made important scientific inroads in illuminating the link between ecosystem function and such services. One of the most important is the delivery of abundant clean water for drinking. Freshwater, while renewable, is nonetheless a resource for which there is no substitute. Humans globally depend heavily on lakes, rivers, and below-ground wells to supply their drinking water. Yet, these primary sources of freshwater contain less than 1 percent of all the water on Earth. The rest is almost entirely saltwater.

Freshwater that is drawn from these sources is replenished by rainfall and snowfall. Rainwater and snowmelt flow overland and back into rivers and lakes as well as penetrating the soil to enter below-ground aquifers. Forests surrounding these water bodies play an important role in the delivery of clean water. By rooting in the soil, trees prevent the soil from being compacted, which allows water to infiltrate and replenish below-ground aquifers. By rooting in the soil, trees prevent soil erosion during run-off events, which prevents river and lake water from becoming murky with suspended soil particles. Such natural "water treatment" helps municipalities offset hundreds of millions of dollars in capital costs of building water treatment facilities. Natural water treatment also offsets annual water filtration costs. A typical American municipality would need to pay about $300,000 per year if 60 percent or more of its watershed remains forested; the cost would escalate with every percentage loss of forest, and it would reach up to $1 million per year if only about 10 percent of its watershed remains forested.

Like most natural rivers, the one in my city's watershed has been left to meander its way through the countryside. The winding ebbs and turns of the river provide an important regulating service. By preventing the flow of water from gathering momentum following heavy rainfall events, the river's structure reduces or even prevents the chance of flash flooding that otherwise could scour the river bottom and heavily erode the riverbank. Scouring homogenizes the river structure, eliminating the distinction between riffles and pools. Riverbank erosion causes the water to become silted and undrinkable. In the course of transforming nature, humans straighten river channels. The intent in the past was to increase the reliability and efficiency of water delivery, but, in many instances, such channelization has done just the opposite. Consequently, the physical integrity of the watershed and of water quality is diminished. Property damage from flooding often occurs.

Channelization eliminates the riffles and pools of the river, structure that provides habitats for numerous species, most notably the many species of aquatic insects that are food for fish. There is a whole industry built around tying artificial flies that emulate these insects in both their immature (technically called larvae, but known in fly fishing vernacular as nymphs) and adult forms. Even while aimlessly browsing through any of the popular fly-fishing supply catalogues it is hard not to stop and be captivated by page upon page of sheer artistry represented in the assorted, realistic-looking insects. They come in all forms, sizes and colors. It is art-imitating life at its best.

The aspect of life imitated by this art is what ecologists call biological diversity or biodiversity. Certainly the different forms, sizes and colors, distinguish different kinds of species. But those forms, sizes and colors also mimic species traits that have functional purposes. These traits help the species to go about their business of living within their habitats.

Take for instance a common group of insect species, the caddisflies, whose larvae live in riffles, wedged between stones in riverbeds. They spin silk catchnets, akin to spider webs, that are then suspended on vegetation emerging from the riverbed. These catchnets are used to capture food, such as algae, detrital particles, and tiny animals that are suspended and drifting in the water column. In doing this, caddisflies provide natural water filtration. The size of the silk catchnet that any given species makes, and the spacing between the silk strands of the woven net (akin to the size of a filter's pores), is determined by the species' body size. The differently sized caddisfly species living together on the river bottom complement each other functionally. They collectively purify the water by removing a wide range of suspended particles. What's more, the species can act like ecosystem engineers that regulate the flow of water. The catchnets, when highly abundant and spread across a river bottom, collectively maintain the river's structural integrity by helping it to resist damages from episodic flooding events.

The trees and shrubs that grow alongside the river also play an important role in aquatic ecosystem nutrition. Most freshwater systems do not have enough nutrients circulating within them to be self-sustaining. Thus, the surrounding terrestrial environment provides an important nutrient subsidy through the countless numbers of dead leaves that fall into the water each year and soil organic matter that leaches into the water. This material provides up to half of the energy and nutrients needed to support the aquatic ecosystem. But the subsidy of leaves and organic matter would likely choke the river if it weren't for other kinds of caddisfly species, as well as a diversity of other species including stoneflies and freshwater shrimp, that break-up this material into tiny pieces. These shredder species certainly consume their needed share of the nutrients during the course of breaking up the leaves. But they also create the smaller suspended detritus particles floating in the water column that are then captured by the net-spinning caddisflies and other filtering species like clams. (Mosquito larvae serve a similar role in ponds.) The wastes released by these decomposer and filtering species become fertilizer for algae that grow on stones on the river bottom. The algae supply nourishment for a diversity of herbivorous insects, crayfish, and fish. In turn, predaceous insects like dragonfly and damselfly larvae, aquatic beetles, and my ultimate quarry, the trout, eat these species. These different kinds of species together operate as a grand nutrient supply chain that builds up the aquatic food web and hence the river's economy.

But things don't stop there. After spending a year or more in the water, the larvae of most of these insect species emerge onto land and develop into flying adults that spend most of their time breeding. The adults (the dry flies in fishing vernacular) are of course food for fish. But the flow of nutrients also comes back full circle to the terrestrial environment, as the adult insects become an important food source for the many songbirds living in the surrounding forest. These birds time their migrations and their life cycles, in part, to coincide with the emergence of the adult insects. During this time of the year, adult insects emerging from rivers can supply up to 60 to 80 percent of the songbirds' diet.

These processes do not just happen in my small stretch of river. They are scalable in space, from managed watercourses in dense urban neighborhoods, to those within larger municipalities or counties, to the grand, untamed rivers in remote wilderness. My favorite stretch of the river may not be remote wilderness in the strict sense. But it nevertheless shares many functional elements and processes of wilderness that one can rely on and value. Scientifically studying the inner workings of ecosystems from the densest urban areas to the most remote wilderness helps us understand the different ways that humans connect to and influence their surrounding environment. It leads to general principles that explain how nature's economy works and the kinds and levels of ecosystem services that humans may expect in those different locations, depending on the degree to which they interact with and influence nature.

Within any ecosystem, services are supplied by literally thousands of species. But, as we have come to learn, there are effectively only a few key functions—like shredding, and filtering, and producing and consuming—that these species carry out. If so many living species provide only a few, similar functions, do we really need to worry about preserving all of them? Surely, losing a few species couldn't be that harmful because there will inevitably be other species that can step up and fill in, right? Would the services provided by nature really change that much if we lost one or a few species?

These are fundamental questions that have occupied the minds and research efforts of ecologists for decades. Ecologists have always been fundamentally occupied with understanding why there is a wide diversity of species to begin with, because it helps to reveal and understand how intricate nature really is. The New Ecology, however, is drawing an ever more scientifically fascinating portrait about how and why the intricate ways that species that have evolved and forged interdependencies with each other matter to sustainability.

* * *

All individuals within species have a simple game plan: contribute as many genetic copies of yourself—offspring—to future generations as you can. Playing this game successfully requires procuring the most resources possible while outwitting your competitors and enemies. Since many strategies can achieve success, different strategies can emerge—evolve—when individuals with different capabilities are pitted against each other. Hence, evolutionary and ecological processes are mutually reinforcing (not separate) processes. The whole of it can be considered an eco-evolutionary game.

To illustrate what I mean, take, as a start, plants. Their strategies involve consuming raw materials like mineral nutrients and water from the soil, and carbon dioxide from the air and, stimulated by the sun's energy, manufacturing plant tissue containing macronutrients: proteins, sugars and starches, and fats. These are the building blocks of all life. Plants allocate these building blocks to make their body parts, like stems, leaves, and roots. Plant traits like stem height and thickness, and degree of branching, determine how well a plant is able to rise upward to gather sunlight. Plant traits like the size, shape, and location of leaves on stems and branches determine how much sunlight can be gathered, and how much water can be retained. Plant traits like root length and thickness, and degree of branching, determine the amount of water and nutrients like nitrogen or phosphorus that can be drawn up from the soil. Plants can vary how much they allocate to build stems, leaves, and roots. Just as one can arrange a basic set of musical notes to create an endless variety of music compositions, plants can arrange their allocations to stems, leaves, and roots to create a dizzying variety of architectural forms and sizes. We use these architectures, or rather collectively distinctive structures, to identify them as species. They use them to maximize their success.

But why are there different architectures to begin with? The answer lies within the evolutionary processes that explain how and why different architectures may ultimately arise.

Not all the individuals within a species build their architecture exactly alike. They develop their stem, leaf, and root traits in different ways dictated to a certain degree by the body plan coded in their genes. Such variety can cause differences in individual capabilities, which in turn leads to differences in success in different environmental conditions.

For instance, some individuals within a plant species might have long, straight roots that penetrate deeply into the soil whereas others might have shorter, branching roots. Individuals with longer roots are likely to grow and reproduce better than shorter-rooted variants whenever soil nutrients and water are found deep down in the soil. Long-rooted individuals will be favored by natural selection because they are better competitors for available soil nutrients. But the advantage would flip to shorter-rooted individuals if nutrients and water occurred closer to the soil surface. Such variety among individuals within a species helps species thrive as environmental conditions change; it provides the capacity for species to be resilient.

Suppose now that long-rooted and short-rooted individuals both tried to occupy sites where the soil nutrients and water occurred mostly at shallow depths. Long-rooted individuals would be at a disadvantage, and thus would become rare or absent there. Short-rooted individuals would, however, thrive and overtake the site. Over the course of time, these individuals could tend to interbreed only with one another. If they passed their genes onto their offspring, then eventually a population of strictly short-rooted individuals could occupy the local site. The population would be specialized—adapted—to draw nutrients from shallow soil layers. If, during that same time they also became less likely or unable to interbreed with their longer-rooted counterparts, we may see the rise of a new species. Thus, variety within a species can also give rise to a diversity of species. And, given that different species may have different forms and sizes, evolution can give rise to greater functional diversity as well. This has important implications for ecosystem functions and services.

Within ecosystems, it is often the case that soil nutrients and water are spread through the soil. It is therefore possible for both long-rooted and shallow-rooted plant species to thrive together as a community because each is adapted to gather its nutrients from a different soil layer. The two are said to occupy different ecological niches. Moreover, nutrients will be gathered in complementary, rather than in interfering or competitive, ways. This is called niche complementarity or functional complementarity. (The collection of caddisfly species with different net sizes is another example of functional complementarity.) Complementarity increases the efficiency with which available soil nutrients are used to build plant-community biomass. Building plant biomass, called primary production, is considered an important ecosystem function; the level of species diversity, expressed as the degree of functional complementarity, can therefore determine the level of ecosystem functioning.

Functional complementarity in rooting depth can also provide important ecosystem services. By penetrating different soil depths, the two species together would maintain soil structure. Together they prevent surface soil erosion that might otherwise occur if short-rooted species weren't present. They also protect the various soil layers and soil porosity that allow water infiltration, which might otherwise not occur if long-rooted species weren't present.

Plant functional complementarity can arise in other ways as well. Individuals with the same rooting depth may also vie for access to sunlight. If some individuals are better able to allocate nutrients to stem height and to large leaves, they could overtop other plants and begin to shade them out. But changing allocations of limiting nutrients forces an economic trade-off. Allocating more to stems and leaves means less is available to allocate to roots. Again, if like individuals tend to interbreed with like individuals, we may see the rise of two species engaged in a different form of spatial complementarity: one that is specialized at gathering sunlight, and another that is specialized at gathering shallow soil nutrients. Individuals with the same rooting depth may also have different timing and rates of life-cycle development. Some may develop earlier in a season, and thus age, reproduce, and die earlier than others. Again, if like individuals tend to interbreed, we may see the rise of two species that live during different times of the growing season, called temporal complementarity. Individuals with the same rooting depth may also have different abilities to use different chemical forms of soil nutrients. Nitrogen, an important soil nutrient, exists in several chemical forms including ammonium and nitrate. Plants might become specialized on one or the other of these forms, giving rise to species that may have the same rooting depth but exhibit complementarity in nutrient use.

These dichotomous sets of traits (long vs. short rooted; tall- vs. short-stemmed; early vs. later seasonality; ammonium-loving vs. nitrate-loving) when combined in different ways can lead to sixteen different architectural forms and life styles of species. All sixteen could complement each other functionally within a given location. In some sense, functional complementarity can be imagined as being akin to what businesses do in a market economy. The different plant species effectively vie for their "market" share in nature's economy, and they thrive by capitalizing on their unique capabilities. A diverse complement of species would use available resources with greater efficiency than a less diverse complement. The consequence is that there will be higher ecosystem productivity than could be realized with a less diverse subset of these species.

In cases of spatial complementarity, plant species may look different due to different architectures that enable them to coexist and to function differently within in a community. But in cases of temporal or nutrient use complementarity, different plant species may still look very similar, even though they too function differently. Ecological science thus shows that it would be folly to triage species by assuming that their similar appearance means that they serve the same purpose within an ecosystem.

The evolution of plant traits has allowed different plant species to cope with different environmental conditions ranging from arid to water-logged, freezing cold to hot and humid, persistently windy to calm, sunny to shady. These adaptations have enabled plants to populate almost every place on Earth and to create the foundation for ecosystems no matter where they take root.

But ecosystems aren't just made up of a diversity of plants. There is a diversity of herbivores that eat those plants. There is a diversity of carnivores that eat the herbivores. There is a diversity of decomposers and microorganisms that break down dead organic matter. Evolutionary diversifying processes can give rise to different species of herbivores, carnivores, decomposers, and microbes. In many cases this involves a game between species, and not just a game among individuals within species. For example, when one player—a prey species—can adapt to fend off its consumer, the consumer in turn adapts countervailing measures. This can lead to tightly interwoven dependencies among species, leading to diversity in the very kinds of functional roles. Importantly, these dependencies are not static; species are perpetually changing and adapting to each other so that they can thrive. It is a process that contributes to the changeability of nature that I alluded to in chapter 1, a process that needs to be preserved if we want ecosystem functions to sustain society.

Consider again plants, but now in association with the herbivores that eat them. Most herbivore species eat stem, leaf, and root tissue. Many insect herbivores also feed on plant sap, plant nectar and pollen, or they mine through leaf tissue. Other insects and nematode worms feed on plant roots. Being even partially eaten can cost plants dearly because it compromises their ability to maximize their own survival and reproduction. As countermeasures, plants have evolved a whole range of defense traits to make them less desirable as foods.

Some plant species change their architecture by producing prickles and thorns; others increase the toughness of their stems and leaves. These strategies make it difficult for herbivores to bite and chew the plant material, and in some cases, as when plant tissue is fortified with silica or lignin, can cause herbivores to wear out their mouth parts. Some herbivore species have countered with mechanical adaptions. The evolution of narrow, nimble mouth parts like those found in forest dwelling African antelope allows them to maneuver around prickles and thorns. The evolution of robust, crushing mouth parts, like those of bison or rhinoceros, allows other species to overcome leaf toughness.

Some plant species try to make themselves indigestible by producing a lot of plant fiber or producing chemicals that reduce the ability of herbivores to digest plant matter by interfering with digestive enzymes and protein absorption. Tannins, which give tea or wine or grapes their bitter taste and dry your mouth of saliva, are but one example. Herbivores respond with behavioral and size adaptations. Some herbivores become generalist feeders. Eating a wide variety of plant species reduces the likelihood that they will consume large concentrations of any one particular defense in any one particular plant species. Often this requires having a large body size to allow the individual to roam widely, as well as having a large gut that can retain plant material for the long time they need to ruminate on it. Other herbivores have become specialized to seek out those rare, small morsels of plant material that are less heavily defended, a strategy that tends to favor small body size.

Some plants are laced with toxins that are deadly in high concentrations. Humans rely on some of those toxins in low dosages of spices that we use in our everyday life, like cinnamon, nutmeg, mace, allspice, mint, pepper, mustard, and chili. Others, including nicotine, caffeine, morphine, cocaine, strychnine, and quinine, may or may not be useful to humans, depending on how one defines useful. These chemicals can cause physiological impairment in herbivores, including interfering with enzyme function, protein and DNA synthesis, cell membrane structure, and nerve transmission. When ingested in high doses, these chemicals can cause any unpleasantness, from nausea and vomiting, to hallucinations and convulsions, and, in the extreme case, death. Most herbivores try to avoid these plants altogether. Others, like caterpillars, have evolved the physiological capacity to eat the plants and store—sequester—the toxins in their own body tissue. Ingeniously, they use these sequestered chemicals in turn to fend off their predators.

Pollen is precious to plants. It is key to producing offspring. Plants will sometimes defend pollen with chemicals. But there is a trade-off, because too much defense may mean that animal species that could transmit pollen from one plant to another don't visit the plant, leading to potential reproductive failure. Thus plants have developed all variety of intricate flower sizes, shapes, and colors that humans deeply admire and cultivate. But, functionally, flowers are intended to attract pollinator species that will collect and transmit pollen to other members of the species. Plant species may each develop unique flower shapes and colors that are tuned to attract one or a few pollinator species to them. Often, nectar rewards are part of the bargain. This way, the plants minimize competing with other plant species for pollinators. This has lead to a diversity of co-evolved plant/pollinator associations in nature. Losing a plant species through triage or disregard therefore has consequences beyond the plant species itself. It may precipitate losses of the pollinator species that are intimately dependent on that plant species. Hence, there will be ripple effects because of the tightly co-evolved interdependencies. Humans have attempted to circumvent these tight interdependencies by relying heavily on the European honeybee as a general pollinator. As we have come to learn, however, this circumvention could have perilous consequences for humans.

Plants also use nectar rewards, or other food rewards, to create interdependencies in other ingenious ways. One in particular is to build food-containing structures on stems or leaves, called extrafloral nectaries, which are intended to attract aggressive species like ants. Ants are nature's bouncers. Their presence scares away any potential insect herbivore from attempting to feed or harassing them while feeding, causing them to give up. But ants will not provide this service if there is no food reward from the plants.

Some plants also use a certain group of chemicals, called volatile compounds, to indirectly fend off herbivores. The volatiles are released as a plume of odor—a "perfume"—into the air whenever insect herbivores start chewing on them and causing damage. This odor attracts flying insect predators and allows them to home in to where the herbivores are feeding on a particular plant. The evolutionary steps that have taken place to attain this interplay baffle imagination (so far).

* * *

Ecologists have made a concerted effort to understand how all of this evolved diversity and the associated dynamism is related to ecosystem functioning. In doing so, they have relied on field experimentation. This definitive scientific method requires researchers to manipulate one or more factors, to measure the outcome, and then to ascribe the measured outcome to what has been manipulated. Experiments that examine diversity/functioning relationships manipulate the diversity of species within a location and then measure the resultant level of ecosystem functioning. To date, such experiments have mostly been conducted in grassland ecosystems using a wide variety of grass and wildflower species.

Grasslands have become important testing grounds for ideas on diversity/functioning relationships. Experiments are done there for reasons of practicality and cost. It is entirely feasible to sow plants into many different field plots and to obtain results fairly quickly within one to several years. Although experimentation in other ecosystems, like forests, or lakes, or oceans is certainly not out of the question, it is often logistically more difficult to experiment in those places. For example, experimenting in forest ecosystems is challenging because of the sheer space and time that would be needed. What is learned only a few years later within several hectares of grasslands would need many, many square kilometers of space, let alone the many, many decades or even centuries that experiments in forests require. But plant species share evolutionary heritages, being formed by similar evolutionary processes. Moreover, the same ecological principles can apply even in widely different systems that have different evolutionary histories, such as coral reefs or tropical forests. Thus, principles that are learned about grasslands are transferrable to other ecosystems. That is, ecologists needn't study exactly every last piece of every corner on the Earth in order to advance well-substantiated working principles about how nature works.

Grassland diversity/functioning experiments typically follow a standard protocol. Diversity is manipulated within plots in fields. Individual plots, ranging from two to four square meters in size and having similar underlying soil properties, are initially sown with either monocultures of native wildflower or of grass species; or they are sown with polyculture ranging up to thirty-two random combinations of wildflower and grass species. Functioning is then measured as plant biomass production over a growing season within a given plot. The experiments are designed to estimate the trend in production or the yield as diversity is increased from monocultures of different species to polycultures of more and increasingly more plant species. The experiments show that polycultures can lead to a doubling or more of yield, compared to the monocultures. That is, plant diversity significantly enhances grassland plant productivity. In part, plants in polycultures make more complete use of available soil nutrients than do monocultures because of the evolved, complementary ways that plant species use resources.

Ecologists are confident in these findings because they are results from many studies in many places. Similar outcomes are observed whether the testing grounds are in Sweden, England, Ireland, Germany, Switzerland, Portugal, Greece, or the United States. Some rare exceptions do still occur, where plant diversity seems to have little or no effect on productivity. In nature, exceptions can arise because of unique characteristics of a field site, such as the plant species available for experiments, or the background soil and environmental conditions. The potential that any single experiment could produce an exceptional outcome underscores the importance of repeating experiments in different locations. Failing to repeat experiments distorts the use of science to solve problems. This is because crafting solutions based on a single experimental test entails a potentially faulty presumption of generality: the solution could be based on a false positive. Moreover, no qualified scientist would apply the findings if the single experiment produced a rare outcome of no effect. But even this could lead to a false negative if after repeated testing we find that the more likely outcome is a beneficial effect. A false negative would then lead to important missed opportunities. Thus, repeating exactly the same experiments in different locations is a necessary precondition for building robust scientific knowledge that minimizes the chance of crafting policy and solutions based on faulty conclusions. Scientists frequently frustrate policy makers by saying "more study needs to be done." It is typically regarded as a thinly veiled ploy to secure more resources to continue yet more research. What scientists mean to say is that more repetition is needed so that we can distinguish likely outcomes from rare exceptions in order to mitigate risks of crafting faulty policy.

Replicated diversity/functioning experiments have also taught us that more diverse systems are more likely to repel invasive species than less diverse systems. Complementarity effectively leads to tightly knit use of space by native plants that repel invasives from appropriating above-ground and below-ground space and thereby getting a foothold.

Experimentation has also shown that less diverse systems are less able to withstand extreme disturbances like drought than more diverse systems. Highly diverse systems of species tend to have a broader portfolio of species with different environmental tolerances than less diverse systems. This greater presence of tolerant species can step up their performance to maintain ecosystem functioning in the face of the disturbance while the performances of other less tolerant species begin to wane.

A species-diverse system in nature's economy is likely better able to mitigate risks of performance loss than a less diverse system in much the same way that a diverse portfolio of stocks and bonds helps mitigate risks in a volatile market economy.

The complementary way that species fit together to maintain the functional integrity of an ecosystem has been likened to rivets on an airplane that fit together to maintain the functional integrity of the airframe. Rivets fasten the metal skin to the airframe and thereby hold together the entire structure so that the plane can fly. The Rivet Hypothesis proposes that one could randomly lose a few rivets (species) without the whole structure (ecosystem) coming apart. But continued loss of rivets would eventually lead to catastrophic failure in functioning.

The Rivet Hypothesis presumes that every species has a unique role to play and thus that each is essential to maintain ecosystem structure and functioning. But most field experiments show that the level of functioning rises and then saturates as species diversity is increased. This means that, beyond a certain level, increasing diversity tends to have diminishing rates of return because many species have similar, rather than unique, functional roles. That is, there is some functional redundancy among species within ecosystems.

Functional redundancy ensures that ecosystem functioning and hence delivery of ecosystem services remain stable in the face of changing environmental conditions. Redundant species are not completely alike: they differ somewhat in their tolerances to different kinds of disturbances. Redundancy provides an insurance service in the sense that there will always be species present that can tolerate a given disturbance and maintain a desired level of ecosystem functionality. Redundancy can also be likened to an investment portfolio in which each stock's (species') performance varies over time. But the variation in performance is not synchronous: some species perform well under some conditions; others perform well under different conditions. The portfolio effect works because there is a collection of species that compensate for each other as environmental conditions change and that as a whole maintain a desired level of ecosystem functioning.

Thus, redundancy helps ecosystems resist loss in functioning when hit by a disturbance. It gives ecosystems the resiliency to rapidly recover from disturbance. It helps ecosystems provide services reliably by dampening variability or volatility in the level of ecosystem functioning.

The idea that functional redundancy provides insurance and portfolio effects is useful for thinking about diversity within a group of species having similar feeding (trophic) relations—say herbivore species or predator species—within an ecosystem. But there is also diversity in the trophic groups (i.e., plants, herbivores, carnivores, detritivores, and others) that make up ecosystems and thereby build out the food chain "scaffolding" (see chapter 2). The combination of species diversity within a trophic group and the diversity of trophic groups within an ecosystem creates an intricately woven network of interdependencies, called a food web. Food webs can be imagined as a network of diverse pathways or conduits for the flow of energy and material. Complex food webs, those with many species with highly interconnected feeding dependencies, have many redundant pathways of energy and nutrient flow. This extent of redundancy also enhances ecosystem stability because it offers detour routes to ensure the reliability of flow to all parts of the ecosystem, should any one species be lost from the system. Anyone who has been stuck in a traffic jam on a major highway because detour routes are unavailable, or has suffered through a lengthy electrical power outage because only one main line connects them to the grid, would appreciate the importance of having a diversity of pathways or conduits of flow.

Analyses of marine, freshwater, and terrestrial food web properties reveal that, despite a diversity of pathways, most of the nutrients and energy end up flowing along a few major routes. At any one time, only a few species within each of the different trophic groups may be dominant in abundance or in their consumption and utilization of energy and nutrients. These species, called strong interactors, are the ones that will have the strongest effects on ecosystem processes. Most other species within the different trophic groups have weak effects, and so are called weak interactors. The strength of species effects are determined experimentally using the protocol described earlier (see chapter 2) to determine the opportunity cost of losing a particular species (in that case it was losing wolves from boreal forest), from an ecosystem. Such experiments require that a focal species is removed from certain experimental plots or locations and that the net ecosystem functioning is then compared to plots or locations where the species is present. Species are deemed strong interactors if there is a large percentage shift in the level of ecosystem functioning in their absence, and the converse holds for weak interactors.

Quantifying the strength of each species in a food web can become quite involved because it requires systematically removing one or more species at a time and doing the comparisons. Such studies provide striking and counterintuitive insight about how species fit together to provide the most stable flow of energy and nutrients. It turns out that the many weakly interacting species operating together can be an important stabilizing force within ecosystems. This is because many weakly interacting species collectively counterbalance the effects of the few strongly interacting species. That is, strongly interacting species are thought to heavily exploit their resources in ways that cause wild fluctuations in ecosystem functioning or services, or that may even collapse the ecosystem. The weakly interacting species importantly counterbalance that effect because they appropriate a share of the resources and thereby offset the "runaway" consumption by the strong interactors.

The lesson here is that humans may need to change how we value species for conservation. Normally we tend to value the abundant and strong interactors within ecosystems because their role is readily apparent to us. Species that are naturally rare within ecosystems are often disregarded because of their comparatively weak effects. But while each of these rare, weakly interacting species work in the shadows of the dominant species they can collectively end up counteracting the effects of the strong interactors. They thereby provide important threads that hold ecosystems together. Conserving the diversity of weak interactors may be as important as conserving the few dominant species to insure reliability and sustainability of ecosystem functions and services. This is akin to market economies where a diversity of large-, mid- and small-capitalized companies that compete for the same market share end up delivering products more reliably and fairly than does a monopoly.

Ultimately, maintaining diversity within ecosystems ensures that a wide range of options is available for adapting to environmental change. So doing ensures that there is a portfolio of species to choose from in order to sustain ecosystem services. It also ensures that ecosystems have the evolutionary capacity from which to select—the genetic variety within species, and the species variety within communities—as conditions change. The next chapter discusses how, by changing this variety, humans can affect ecosystems.

CHAPTER 4

DOMESTICATED NATURE

After being away for decades, from a place that had had a profound, formative influence on me, I had occasion to return to the small forest and adjacent meadow bordering a creek at the edge of the small town in central Ontario, where I grew up. I belong to that bygone generation whose parents still encouraged their children to spend entire days unsupervised, freely exploring their environs. It was a childhood that Richard Louv wistfully reminisces in his book, *Last Child in the Woods*. I could go to this special place any time of day, during any season, and become immersed in it: breathing and smelling its air, listening, observing, touching, and tasting to discover its wonders.

The place was home to a diversity of bird species that occupied habitat spaces that complemented one another. I discovered that some species like whippoorwills, woodcock, and grouse preferred to live on the ground; other species like thrushes and flycatchers lived throughout the lower-level shrubs; others like warblers lived in the upper canopy. They had different coloration to blend in with the part of the habitat where they lived. They had different beak shapes and sizes to deal with the different foods that they ate. I watched many of them flitting back and forth between the forest and the meadow. The meadow was always teeming with butterflies and bees and grasshoppers and beetles. I observed their development during entire summers. I discovered how insect life cycles worked, and how they were precisely timed with the changes in the growth and development of different wildflower species. At the forest/field margin I saw my first pileated woodpecker, which at twenty inches in length is a giant among all the North American woodpecker species. I encountered my first snowshoe hare there and learned first-hand that it changes its coat color from white in winter to grizzled brown throughout the rest of the year. I encountered my first red fox there and marveled at its agility when hunting mice. I learned to read animal tracks in the snow and decipher which species made them. Gaining this rich natural history knowledge was like doing an apprenticeship of sorts. It kindled my interest in learning much more. It prepared me to undertake formal study of how nature works. It is why I chose to become an ecologist.

So it came as quite a shock when I returned after twenty years to find that the place had been completely transformed. The creek had been dammed and the entire field and surrounding forest was flooded by a large water reservoir. Many of the once verdant trees had been cut down; the rest were now mere standing deadwood in the water. The species that were so familiar to me were nowhere to be found.

Such conversion of land can also do a lot to change ecosystem functioning, especially the balance of carbon and nutrients entering and exiting the system. Cutting down trees and submersing the land means there is less runoff of organic matter to support the aquatic food chain—on the order of a quarter as much as would normally enter the stream. The organic matter that does enter the reservoir remains suspended in the water column because many of the shredding and filtering species that reside in streams become lost from the system. This in turn lowers algal production, which lowers the amount of carbon and nitrogen taken up and cycled within in the aquatic food chain. Instead, bacteria living in the water column decompose the organic matter. This bacterial action can cause about three times more carbon dioxide per unit area to be released back to the atmosphere from the reservoir, compared to what the stream would have released. The bacterial action also releases methane, which is a far more potent greenhouse gas than is carbon dioxide.

One could become outraged and wonder why anyone would do this to this wonderful place. But then, how could one be angry at a colony of beavers? They simply went on fulfilling their naturally evolved role as ecosystem engineers that transform environments to suit their own needs. In fact, the pond reservoir was also teeming with life, even though not the species that I was used to. The meadow wildflowers were replaced with water-loving species like sedges, marsh marigolds, cattails, and pond lilies. All sorts of water birds including ducks, herons, kingfishers, and red-winged blackbirds now lived there. The terrestrial insects were replaced by aquatic species including water striders and water bugs, mayflies and dragonflies, and whirligig and diving beetles. And of course the pond now had its resident frogs and small painted turtles. By altering the physical structure of the area, the beaver created an environment that built up a new food chain—and thereby a new natural economy—by attracting an entirely different collection of plant, herbivore, carnivore, and decomposer species. The beaver colony effectively controlled and exploited nature to suit its own needs.

Beaver are often the first species among many that come to mind when thinking about ecosystem engineers. But termites, in particular, can do things on even grander scales. Granted, termites have a notorious reputation as the bane of urban environments. We normally would stop at nothing to exterminate them to prevent the harm they cause to anything in their sight that is made of wood. In places like Africa, however, termites are responsible for the spatial patterning of savanna ecosystems and are therefore beneficial to a large diversity of plant and animal species that exist there.

The spatial patterning arises from what ecologists call a self-organizing process. That is, there is no overarching, coordinating force that structures the landscape. The spatial patterning emerges more spontaneously as the termites go about their business of living.

Termites, for example, live in large colonies that extend from a meter or more underground to several meters aboveground. Made of a giant earthen mound, a colony's home resembles an ancient, dead tree trunk. The mounds in fact are built to be natural air conditioners that circulate and exchange air to maintain steady year-round environmental conditions within the colony's home. The mound also becomes the center of the colony's foraging territory. Termite feeding forays from this central place are eventually inhibited when individual termites encounter and compete with members of adjacent colonies. Amazingly, this behavior can lead to quite regular spacing of termite colonies across the landscape.

The soil close to the mounds also becomes sandier than areas many meters away, and this sandy area becomes grass-covered. The sandy soil promotes water infiltration, aeration, and nutrient build-up around the mound sites. A bird's-eye view of the landscape reveals that miles and miles of evenly spaced termite mounds create evenly spaced moisture and nutrient "oases" within what would otherwise be a parched landscape. Concentrated moisture and nutrients in turn promote tree growth that then fosters the build-up of food chains comprised of insect herbivores and spider and lizard predators of the insects. Termite colonies build up new natural economies by promoting nutrient supply for primary production, which in turn supports secondary and

tertiary production as nutrients flow to herbivores and then to predators. The termites even farm fungi within their colonies to decompose wood fiber and thereby release nutrients that they in turn consume. The released nutrients also nourish plants growing in the vicinity of the mound, contributing to the productivity hotspots.

These productivity hotspots that emerge evenly across the landscape also attract and support numerous, widely ranging large mammal grazers like zebra, buffalo, white rhino, impala, and blue wildebeest; and large mammal browsers like kudu and giraffe. Of course, large predators like lions and leopards follow.

Ecological engineers are major agents of change. They invade new environments and transform them to suit their own needs. They routinely destroy one ecosystem and its associated natural economy only to build up another one. In this sense, humans and their actions are not an exception. Like other ecosystem engineering species, humans are merely expressing naturally evolved tendencies to transform their environments. Humans, however, are exceptional in the extent to which they re-engineer the world to suit their own needs, a process that has been dubbed domesticating nature. The extent to which humans are domesticating nature can be cause for alarm on many fronts. It raises concern about human disregard for the evolutionary history and intrinsic value of life on Earth. It raises concern about the threat to ecological functions and services that support humankind. The New Ecology is grappling with the issue of what an increasingly domesticated world means for the inner workings of nature. Ecological scientific study is already revealing important impacts on ecosystems that could jeopardize sustainability. But in doing so, it is also newly uncovering interesting ways that species may respond to changes wrought by humans. Human domestication of nature presents important challenges to the adequacy of many traditional ecological theories to inform as well as to apply ecological scientific enquiry to solve environmental problems. Ecologists have responded by beginning to advance new theories about how nature works.

* * *

Humans engineer landscapes to support their demands for food, housing, transportation, energy, and technology. Globally, landscapes have been transformed by enterprises like agriculture, logging, mining, and construction. Indeed, about 50 percent of the Earth's land surface has been converted for crop and livestock production alone, with half of the world's forested landscapes repurposed to that single land use.

Like termites, human engineering of agricultural land effectively creates hotspots of productivity. Crops, like other plant species, rely on the creation and maintenance of highly fertile soils, and of abundant and diverse soil biota. Crop production creates opportunities to build up new food chains by attracting insect herbivores. But humans treat these species as pests because they consume that production. Enlisting the ecosystem services of predaceous natural enemies can control crop damage by insect pests. By applying the ecological principle of trophic cascade, ecologists can attract or release populations of predaceous natural enemies into the crop fields to attack the insects and thereby reduce their impact on crop plants. However, such measures have variable success rates. Success depends on the availability of habitat for natural enemies in the landscape around the agricultural fields. Success also depends on whether the traits of the predators are suited to dispatching particular pest species. Finding suitable predators oftentimes requires lengthy trial and error testing. So, in the interest of expediency and reliability, human engineering often circumvents altogether the use of natural enemies in favor of applying chemical pesticides or growing genetically modified crops that resist herbivores in the first place.

As with other species of engineers, the economy arising from land transformation to crop production is created to meet the needs of the species that engineered it. But, unlike other species, such human transformation is expressly geared to steer all that primary productivity to just one species. By expanding the lands dedicated to this enterprise, and further by exploiting forests for timber and fuel wood, humans withdraw and divert productivity away from food chains, thereby supporting less biotic diversity. By eliminating unwanted species that vie for a share of the production, domestication of land for crop production actively tears down food chains, rather than building up new ones. It further homogenizes species diversity by cultivating monocultures of a handful of selected crops.

Ecologists have calculated that humans appropriate about 25 to 30 percent of the Earth's potential primary production for their exclusive use. These numbers are based on the estimated difference between plant productivity on Earth that would be available if human land transformation and exploitation did not occur at all, and the amount that is left over under current conditions of domestication.

Such calculations are made possible by using state-of-the-art Earth-observing satellite technology. Satellites are used to take time-elapsed snapshots of the entire Earth remotely from outer space. The snapshots reveal the different land uses across the globe. Remote sensing of the Earth also allows rapid assessment of the condition of vegetated areas across the globe. Condition is based on an index—the Normalized Difference Vegetation Index (NDVI)—that is calibrated to different wavelengths of light used by different types of plants during photosynthesis. The index can be used to tell whether the primary production is from forests or grasslands or other types of vegetated lands or whether it's from aquatic plants in oceans or lakes. The index pinpoints places across the globe where plants are highly productive, less productive, or not productive at all; and whether the productivity is due to land clearing or land transformation. Comparison of the snapshots reveals the time course of change in land use, where it is happening the most, and where it may be headed.

Assuming that the current pace of change will remain unabated, ecologists estimate that humans stand to appropriate about 45 percent of global primary production by the middle of the twenty-first century. Much of it is used to feed the burgeoning global human population and livestock; or used as biofuels, firewood and building materials. This feat of ecosystem engineering and domestication, and its global reach, is even more remarkable considering that humans comprise only one-half of 1 percent of the total biomass of all the species on Earth that need primary production in one form or another to sustain themselves.

As consumers, humans also appropriate primary and secondary production from non-agricultural ecosystems by harvesting wild populations from all four functional groups—plants, herbivores, carnivores, and decomposers—of food webs. Humans harvest wild plant products like berries, nuts, and grains; and from decomposer species like mushrooms. Commercial and recreational fisheries typically harvest predator species (e.g., tuna, swordfish, salmon, trout). Game hunting harvests herbivore species (e.g., moose, deer, elk). Humans also appropriate animal (secondary) production for themselves by replacing rangeland gazers like bison, deer, antelope, and kangaroos, with cattle and sheep. Predator control measures that are intended to protect cattle and sheep eliminate competitors for that secondary productivity and thereby shorten food chains. Shortening food chains alters nutrient and energy flows on rangelands.

Human harvests of wild populations can cause evolutionary changes, instigated by the tendency to select individuals having certain sizes, morphologies, or behaviors. Humanly caused evolutionary change in species is estimated to be up to three times greater than the natural, background evolutionary change that would occur in those wild populations. And the changes are rapid, many being within twenty to forty years: less than a human lifetime.

The most visible changes come from trophy hunting, which selects individuals having the largest antlers or horns or tusks. The sizes of antlers, horns, and tusks are traits that are passed on genetically from parents to offspring. Trophy hunting thereafter leads to populations comprised of male individuals having smaller antlers, horns, or tusks. The size of these male traits in part determines their breeding success. The traits are related to the abilities of males to successfully compete for access to females. They also influence the likelihood that females would be willing to mate with them in the first place. Rapid evolutionary changes in morphological features have insidious consequences by altering the breeding systems and ultimately the genetic makeup of the wild populations. This ends up lowering average population reproductive rates. Ultimately, trophy hunting jeopardizes the long-term sustainability of the population harvests.

Rapid evolutionary changes have also been documented in commercially harvested fish, including stocks of Atlantic Ocean and North Sea cod, plaice, halibut, and flounder, and Pacific Ocean salmon. In fish, the number of eggs produced is strongly related to body size. But reaching large sizes requires lengthy periods of growth and development as individuals mature to adulthood. Commercial fisheries selectively remove the largest individuals and thus remove adults with higher potential breeding output. Harvesting ends up leaving stocks mostly if not entirely comprised of individuals whose genetic makeup makes them mature more rapidly, but reach smaller adult sizes. Again, the insidious consequence is a decline in the breeding structure of the population, which results in lower reproductive output of the harvested stocks. Commercial fishing instigates rapid evolutionary changes that may result in increasingly less biomass of fish that can be sustainably harvested.

Fisheries management often tries to reverse declines in harvestable biomass by instituting moratoria on commercial fish harvests. Such moratoria have considerable social and economic costs to commercial fishing enterprises and their associated communities. And yet, the moratoria may not lead to a rebound if individuals with the genetic makeup to mature later and to reach larger sizes are no longer part of a population. Selective harvesting can thus have long-lasting effects if it homogenizes [genetic] diversity in ways that reduce the resiliency of the harvested stocks. The collapse of the northern cod fishery in northeastern North America is a striking example of the costs of losing genetic resiliency in harvested populations. Despite a moratorium on fishing, which had been in place since the mid 1990s, there are little if any signs to this day that the cod stocks are rebounding to levels that can resurrect a commercial fishery.

Curiously, harvests of wild populations by humans—which select out the most productive individuals and thereby deteriorate breeding stocks—operate in a strikingly different way than such selection works with agriculture. Agriculture, for millennia, has kept the most productive individuals of animals and plants in order to enhance breeding stocks. This enhancement ensures that cattle herds are healthy with high outputs of milk or meat production, and that crops are able to produce the highest yields in a variety of growing conditions.

Human activities like logging, mining, and development of urban infrastructure and of transportation networks, in addition to agricultural development, can lead to other evolutionary changes that have associated ripple effects for ecosystem functioning. Most notable is fragmentation of habitat, in which large contiguous habitats become cut into smaller parcels that are isolated from each other. The size of the remnant patches as well as the extent to which they are isolated from each other determines how much biodiversity persists on landscapes. Shrinking habitat size means there is less living space. Inevitably, fewer species live on the fragmented landscape space than existed when it was still covered by intact, contiguous habitat. Compounding the loss of habitat is patch isolation. Increasing the distance between remnant patches makes it harder for species to migrate around the landscape in order to find and colonize new places with suitable living space.

The suburbanization of chaparral habitats in southern California is an illustrative case in point. The human desire to move outside larger urban centers, to enjoy some semblance of nature, has led to the conversion of large swaths of chaparral habitat into residential housing developments interspersed among remnant parcels of the original habitat. Chaparral habitats are normally home to large predators like coyotes. But the remnant parcels are too small for coyotes to live exclusively within them: they must thereupon wander from one patch to another to find prey. Coyotes are quite opportunistic species. Their traits serve them well by allowing them to be unfettered by human disturbance. They move freely across humanly dominated landscapes and indeed can thrive in these new environments. But this also causes them trouble because they come into conflict with humans whenever they prey on house pets whose owners have let them roam unsupervised. Humans respond by exterminating the coyotes. But habitat fragmentation and coyote extermination have knock-on effects. The decline in this large predator has resulted in decreased control over populations of mid-sized predators like foxes, skunks, and raccoons that thrive within the remnant patches—a phenomenon called meso-predator release. These mid-sized predator species prey on songbirds, including their eggs and their young. Meso-predator release has precipitated declines in the diversity and abundance of resident songbirds that are otherwise capable of thriving in the patches. Fragmentation in the California chaparral has unwittingly altered the original natural economy by causing system-wide feedbacks that end up restructuring food webs and how they function.

Loss of diversity and restructuring of food webs are typical consequences of fragmentation. Ecologists know this from deploying decades-long, large-scale experiments in tropical forests in Brazil and Borneo, in temperate forests in Australia and the United States, in prairie grasslands in the United States, and in moss habitats in Canada and the United Kingdom. Ecologists have also at the same time documented their effects on birds, mammals, and insects. These experiments covered areas ranging from 100 to 1,000 square kilometers of landscape space. They systematically transformed contiguous habitat into patches of different sizes, and at the same time thereby isolating those patches at different distances from each other. Most importantly, they included companion landscape spaces that were left intact (as experimental controls) in order to simultaneously compare the effects of fragmentation with similar-sized "patches" within unaltered areas.

The experimental evidence shows that isolating and shrinking the sizes of habitat patches, on average, caused a 30 percent decline in the number of species that resided within the remnant patches. The declines occurred in all trophic levels. Habitat fragmentation also created much flux because some species went locally extinct, to be replaced by new species that colonized the patches. All in all, this meant that the long-term likelihood that any one species persisted in a patch declined by 80 percent. At the same time, the strength of food chain interactions and the amount of nutrients retained within the patches declined by 30 to 50 percent. Increasing the degree to which patches were isolated caused similar changes, including reducing the movement of species between patches by 70 percent, thereby lowering the chance that any one species would recolonize habitats.

Layered upon all of this, with potentially conflating effects, is global climate change. Domestication of nature by humans increases greenhouse gas emissions through land clearing and resource exploitation, land conversion for agriculture, rearing livestock, production and use of cement for infrastructure development, energy generation, and transportation of humans, their goods, and their materials. A warming Earth selects for those species with the suite of physiological traits that allow them to adapt to changing conditions. Those that are incapable go extinct. But the coping mechanisms may vary among species. Some may have the physiological capacity to tolerate or adapt to warming *in situ*; others may adapt by migrating to thermally more favorable locations. Of course, this assumes that the human-built environment maintains ample habitat for viable populations of species to remain *in situ*, or otherwise does not impede species migration across landscapes. And, not the least, all this reshuffling of species across landscapes stands to further restructure food web dependencies and to alter ecosystem functioning, especially so whenever species belonging to certain food webs migrate at different rates.

I am torn by the change wrought by the human enterprise. My childhood apprenticeship in natural history shaped the values, the ethic, and the sense of wonder that I now inescapably hold. I have a deep and abiding respect for the whole diversity of living organisms, their habits, and their habitats. My childhood fascination with nature has endured. It is what keeps me asking the probing questions that let me learn scientifically how things fit together to build up sustainable natural economies. It thus saddens—sometime even maddens—me to see the loss of species, their habits, and their habitats, often willfully, in the name of human "progress." This is because, as an ecologist, I see the hallmarks of change that could irrevocably erase our amazing evolutionary heritage as well as reduce the capacity to sustain all life on Earth. As a scientist, however, I have to admit that I find these changes to be rather fascinating. They force me to see and appreciate the dynamism of nature from fundamentally new vantage points. It lays bare the inadequacy of our classic, textbook principles when what is at issue is how sustainability could be achieved in an increasingly domesticated world. The rapid pace of change has presented new opportunities for scientific discoveries. It has led to a sea-change in ecological scientific world views and thereby demanded the development of new theories and knowledge to better support efforts to achieve sustainability in a domesticated world.

* * *

A classic paradigm of ecology is that ecosystems are self-contained, self-supporting systems. This world view holds that an ecosystem's functions, like production and nutrient cycling, are basically sustained by the species residing within its boundaries. Anything that happens outside the boundaries, including changes caused by humans, is deemed irrelevant to the ecosystem's inner workings. But ecosystems, especially aquatic ones, are not completely self-contained and self-supporting economies (see chapter 3). Seasonal organic matter and nutrient inputs from terrestrial snowmelt and rainwater runoff, for example, routinely subsidize lake and river economies. Wind carries seeds and insects across boundaries. But the classic view tended to dismiss the potential for such external inputs, or from other forms of disturbance like extreme weather events, to have lasting effects.

They were merely considered temporary blips whose effects would eventually be dampened down by strong species interactions within the ecosystems. It was held that this dampening, or feedback, played a key role in sustaining the species composition and thereby in recovering normal levels of functioning. That is, ecosystems always returned to their inherent, internal balance—a balance of nature—in which things were steadied and thereby sustained during the long run.

The notion of self-contained, self-supporting system is implicit in well-intentioned conservation actions that placed fixed political boundaries around prized pieces of wilderness. These places, which we call parks and protected areas, are designed to keep humans and their impacts out (or at least to regulate their access), and to keep the native flora and fauna in. This way, the thinking goes, the balanced state of species and associated natural ecosystem functioning would be protected in perpetuity. And the larger the protected area, the more likely it is to maintain high levels of diversity and functioning. For the longest time in the history of ecology, therefore, the research mantra was: look for places like this, protected from human domestication, to study how nature works. This implicitly promoted the world view of a human/nature divide.

It also encouraged a trend in which human land use and land conversion encroached up to the very edge of protected areas. This has created sharp boundaries that can sometimes be seen even from outer space.[1] Many parks and protected areas effectively became isolated wilderness "islands" in a sea of domesticate land uses.

At the same time, ecologists like Gary Polis, a professor at the University of California at Davis who studied how food webs were regulated, began to question the notion that ecosystems are self-contained when the facts from his own research on oceanic island ecosystems off Baja California did not conform to the prevailing theory. For example, oceanic island ecosystems are sharply separated from each other and from the mainland by large distances and a seemingly impermeable saltwater barrier. These inhospitably arid islands, being mostly covered by *Opuntia* cactus, had low primary productivity. Herbivorous insect species were accordingly rare. The islands nevertheless supported extraordinarily high abundances of spider predators; more so on smaller than on larger islands. This didn't make sense in light of reigning ecological principles. Larger island and intact mainland ecosystems should be more likely to support longer food chains and greater abundances because they are more productive and provide more living space.

Polis and colleagues discovered that island ecosystems are not self-contained, self-supporting entities and that the internal working of an island ecosystem closely involves the size of the island and what flows across its boundary. Instead of concentrating solely

1 http://www.usgs.gov/blogs/features/usgs_top_story/space-veteran-landsat-7-marks-15-years-of-observing-earth/.

within the island ecosystem, they focused on the flow of nutrients and materials across the island boundary. This led to the discovery that the boundary of the island—the shoreline—was not impermeable. Dead algae and drowned animal carcasses that washed up onto the shore from oceanic drift provided important subsidies for the island economies. The shoreline sustained high abundances of detritivorous insect species that consumed the algae and scavenged the decomposing carcasses. The detritivorous insects became an abundant resource that sustained the predatory spiders and scorpions. The detritus subsidy from the ocean propped up the island economy.

The smaller islands supported more predators because of their physical properties. Smaller islands have a higher ratio of perimeter to area, meaning that they have more shoreline relative to their overall area than do larger islands. This allows consumers from all over the small island to access the subsidy. By contrast, individuals living in the middle of the larger islands have a lower likelihood of encountering the subsidy. That is, the heterogeneous spatial arrangement of species and resources across the islands mattered considerably. This challenged another classical view: that resources and species are evenly—homogeneously—arranged in space within self-contained ecosystems.

The abnormally high abundance of predators led to a feedback in which the abundant predator trophic level controlled the abundance on the island's herbivorous insects, thereby further contributing to their rarity on the islands. This lessens the insect damage to plants. Hence the effects of the subsidy reverberate throughout the entire island food web.

The lesson here is that two very different kinds of ecosystems can be inextricably linked through resource flows across their boundaries. The amount of subsidy provided and its attendant ecosystem-wide effects depend on the spatial arrangement of the donor and recipient ecosystems and on the species interactions within both of them. If, for example, marine production is altered by environmental impacts, or from species imbalances in the marine food chain due to, say, overfishing, then the amounts of algae and animal carcasses that subsidize the island economies become altered too. Shut off the marine subsidy completely and the island ecosystem is likely to collapse to a barren desert.

This work also showed how such important insights can sometimes be hard-earned because doing field research can be perilous. Sadly, Polis and several colleagues drowned when, on their way to sample the islands, a freak windstorm capsized their boat.

Paradigm changing discoveries like this sometimes elude scientists until technological advances can abet them to see things differently. Polis and colleagues were able to prove that ecosystems were not self-contained because of the development of modern stable isotope analysis and ready access to mass spectrometers—high technology instruments—that can assay the isotopic content of ecological samples. Isotopes are forms of the same chemical element (e.g., carbon or nitrogen or phosphorus) that have different atomic masses and therefore have slightly different chemical and physical properties. The different isotopic forms of elements co-occur in nature. But organisms preferentially metabolize the lighter forms because it is cheaper metabolically. Organisms thus accumulate heavier forms of, say, carbon or nitrogen isotopes in their body as they consume their resources. Accordingly, organisms higher up the food chain tend to have more distorted ratios of the lighter than of the heavier forms of elements in their tissue than do organisms lower down in the food chain. Different ecosystems (e.g., marine vs. terrestrial) also have different isotopic signatures determined by the ratios of the heavier to the lighter elemental forms available in those environments. It is therefore possible to trace the ultimate food source of an organism based on isotope ratios in its body in relation to the isotope ratios of candidate food resources (e.g., marine-derived algae, or animal carcasses vs. *Opuntia* cactus).

Isotope analysis has led to the widespread discovery that it is the ebb and flow of resources across ecosystem boundaries that drives nature's economies as much as the internal workings within the ecosystems themselves. The ebb and flow can involve organic matter inputs, such as those from terrestrial runoff into streams and lakes or from washing up on Baja islands. It can also result from animals physically translocating nutrients across boundaries. This happens when insects emerging from streams become a major food source for songbirds in adjoining forests, or when fish-eating seabirds release guano onto islands where they gather into breeding colonies, or when salmon migrating to places like Bristol Bay release ocean-derived nutrients contained in their body tissue into their natal streams after they breed and die, or when grizzly bears capture some of those salmon and carry them upslope to be partially eaten and eventually released onto the forest floor by the grizzlies as body wastes. Such nutrient translocation can provide substantial subsidies to recipient ecosystems. In the latter case, isotopic analyses have shown that up to 25 percent of the upslope forest's nitrogen supply may originate in salmon carcasses.

The discovery of cross-ecosystem nutrient flows showed that it is no longer tenable to think of ecosystems as self-contained entities. There is effectively no inherent and steady internal balance. Instead, nature is rambunctious: nutrients are perpetually redistributed and organisms make necessary readjustments in their behavior, physiology, and life cycles in response to nutrient supply in time and space.

The specter of habitat fragmentation also calls into question the classical principle of self-contained systems, especially as it influences how species interact with each other. Here again, the island metaphor is an apt way to begin thinking anew about how biodiversity across landscapes will be influenced by domestication.

Classic theory maintains that ecological communities are held together in large contiguous habitats by competitive or predatory interactions among species. Species coexist because their morphology, physiology, and behavior became adapted to cope with each other's competitive or predatory pressures. This in turn enabled species to divide up their habitat or food resources through the process of niche complementarity (see chapter 3). Ecological systems were said to remain in balance because each species fit into the community according to its specific adaptations. For example, the Serengeti ecosystem supports upward of ten mammalian predators ranging in size from the largest (lions and hyenas) to intermediate (leopards and cheetahs) to the smallest (wild cats and jackals). The size of the predators determines the size of prey that they can efficiently dispatch. But, as a community, the predators fit together in a nested way. The smallest predators can capture and subdue only the smallest size prey. Larger predators can take progressively wider ranges of prey size. The largest predator—the lion—utilizes the fullest range of prey, and so its competitive advantage tends to dominate the predator community: it is the king of the beasts.

This view allows that some species might dominate communities because of their strong competitive effects over more subordinate species. Species can then be ranked according to their competitive ability to dominate a community, which also tends to determine a species' abundance within a community. Species with lowest dominance rankings likely have the hardest time vying for their share of habitat space or resources. These become naturally rare species that often eke out a living by wandering here and there to gain a temporary foothold until such time as they are squeezed out by a higher-ranked species. They are effectively vagrants: their success depends on their penchant to migrate around the landscape. The dominance ranking therefore also represents a reverse ranking in dispersal ability. The least dominant species tend to be best adapted to disperse around the landscape; the most dominant are the poorest dispersers. In between these extremes are species with intermediate abilities to dominate and disperse.

Habitat fragmentation can turn the game entirely on its head. Species that were once spread continuously across the landscape become divided into subpopulations that are now largely confined to local habitat patches. At any one time, a habitat patch may become empty. This can happen if harsh environmental conditions or genetic factors like inbreeding cause crashes in survival or in the reproductive ability of the patch occupants. Migrants from other patches might re-colonize those empty patches.

Fragmented landscapes stand to change the species makeup within and among patches. The better dispersers will likely seek out vacant patches first. Whether subsequent migrant species colonize a patch, or are repelled, hinges on their dominance ranking relative to prior arrivals that have established a beachhead. The landscape will be in continual flux, with species migrating between patches and species composition turning over within patches. Therefore, no two patches are guaranteed to have the same species composition: there is no single, steady balance of species.

Habitat fragmentation also may not have an immediate, noticeable effect on biodiversity. Delays in extinction may occur because the large population numbers of the dominant species may dwindle only slowly as fragmentation and habitat loss progresses. This may reinforce a perception that fragmentation is having minimal impacts, which encourages further land transformation. Continued land transformation and habitat shrinkage may, however, reach a threshold beyond which species are put on a track of irreversible loss. That is, society may create what is called an extinction debt. This debt will be inherited by future generations. In such a case, future generations would be consigned to inherit a landscape in which they can only watch the disappearance of cherished species.

Although this new view of the dynamism of landscapes highlights the potential for extinction debt, it also offers guidance to help offset the likelihood that extinction will occur. Ecological science shows that larger fragments of habitat should have lower rates of species loss than smaller fragments have. Fragments spaced closer together should facilitate re-colonization better than can habitat patches spaced farther apart. Using corridors to keep habitat patches connected, or to reconnect them, encourages species movement. Connecting habitat patches especially enhances the prospects of the poorer-dispersing species that would otherwise be at the highest risk of extinction.

This knowledge in combination with modern technology can assist in thoughtfully planning out how habitat patches might be configured across landscapes by pinpointing critical areas that support species movements. This technology includes, again, satellite imaging of the Earth to characterize in precise detail the spatial arrangement of landforms, habitats, and land uses across landscapes. Global positioning satellite (GPS) telemetry can pick up signals transmitted from animals fitted with radiowave transmitting collars that precisely record their locations on the landscape. All the data can be entered into advanced computer analysis systems (geographic information systems [GIS]) that integrate all of the data to develop comprehensive maps depicting the state-of-play across landscapes. Such analyses can reveal habitat locations where species tend to reside. Also by depicting so-called resistance surfaces, such analyses can show where species flow across landscapes freely and also where their movements are constrained or hindered by natural or artificial impediments and barriers. It helps reveal which places need to be reconnected (parks and currently protected areas as well as unprotected habitat) in order to avert extinctions.

Resistance surfaces provide strategic information that can guide land use planning for sustainable development of agriculture, energy, transportation, urban infrastructure, and conservation. Such information helps overcome the human/nature divide by showing potential ways that landscapes could be configured to allow humans and their enterprises to coexist alongside other species and the ecosystems services they provide.

Configuring landscapes to overcome the human/nature divide requires that species have the capacity to adapt and thrive in these newly created, never-before-seen domesticated landscapes. For the longest time, it was believed that species could not adapt to these growing human pressures. This thinking stems from the classic view in ecology, which held that ecological and evolutionary processes operated separately and on entirely different time scales. Evolution was said to be a slow process, and the biodiversity that we see today was considered to be the product of a process that had happened in the past and required millennia to reveal the species we see today. According to this view, the traits that species have adapted over the course of their evolutionary history are the very traits that now impose significant constraints on what those species can do going forward in time. That is, species adaptations pre-determine their current functional roles and accordingly how ecological processes unfold during future decades extending to centuries. This view leads to the perception that most species are incapable of adapting fast enough to the rapid environmental changes that humans are now creating as they domesticate nature. It is what leads many ecologists and conservationists to worry about massive extinctions of species.

Ironically, it is humanly caused changes that have been instrumental in revealing that the pace of evolutionary change can be quite rapid. Indeed, evolutionary and ecological processes may even operate contemporaneously. For example, harvesting wild populations can instigate evolutionary changes during mere decades. There are other notable examples. Agricultural pest control can quickly lead to pesticide-resistant populations of insects. Damming rivers can cut off migrations by fish species like steelhead trout or alewives that travel from freshwater to ocean and back again during the course of their lifetime. In just 100 years, land-locked populations have become locally adapted to cope with year-round freshwater conditions. These populations have morphologies, life cycles and reproductive patterns different from populations that are still free to migrate, so much so that individuals from land-locked and migratory populations may no longer be capable of interbreeding. This constitutes evidence for rapid speciation.

These revelations have led ecologists to reconsider classical ideas and to probe more deeply to understand the mechanism driving the systems they study. The case of the Galapagos finches is a widely celebrated demonstration of rapid evolutionary change. But insight comes from many other studies as well. Collectively, they are helping us to develop a fascinating portrait of the interplay between the rapid adaptive capacities of species and their ecological consequences.

Anolis lizards are a group of species that inhabit countless tiny islands in the Caribbean. As a clear example of evolution leading to niche complementarity, these arthropod-eating species occupy many different locations within their habitat: some are ground-dwelling; others live on trunks of bushes; and others live only on branches. One can readily discern which habitat locations each species uses based on body and limb morphology. But experimentation has shown that these traits can be quite malleable as the environment changes.

The experiment involved several, small postage-stamp islands that were inhabited by a single species of ground-dwelling *Anolis*. An exclusively ground-dwelling species of lizard that was an effective predator of the *Anolis* was introduced to half of the islands; the remaining islands were left as unmanipulated experimental controls. The predatory lizard did two things, one quickly, the other taking a bit longer. It quickly dispatched individual Anoles that were poorly capable of escaping by climbing on trunks and in branches. During the season, it also caused developmental changes in limb morphology of the survivors, which allowed them to more nimbly walk on thin branches and catch prey in the higher vegetation. Such developmental change during the course of an individual's lifetime is termed phenotypic plasticity. The propensity to undergo plastic changes is however, genetically determined. Some individuals are therefore better able to undergo plastic changes than others. The ones with poorer abilities to change have poorer survival on thin branches in the canopy. They must either try to eke out a living as best they can in the available vegetation or move down toward the ground where they are better able to catch prey. Moving lower down of course risks being caught by the predatory lizard. Either way, those individuals will have lower survival than will the more behaviorally and morphologically plastic individuals. Moreover, the plastic individuals, being the ones mostly remaining in the population, will tend to interbreed with one another. They will thus produce generations of offspring that are more likely to develop the morphology that allows them to live on branches in the vegetation. Thus, phenotypic plasticity has enabled individuals to improve their immediate survival in response to a rapid environmental change. This plasticity has further set in motion longer-term adaptive evolutionary change in which individuals that develop the morphology to thrive in the upper vegetation dominate the genetic makeup of the population. Such changes were not at all observed on the unmanipulated control islands. This experiment thus provides strong evidence that, in very short order, evolutionary processes are causing a once exclusively ground-dwelling species to become locally adapted to the new and unique island conditions.

Guppies—the small, brightly colored fish species of aquarium lore—offer additional insight into the consequences of rapid evolutionary change in response to natural predation. A study conducted in their native streams in Trinidad found that guppy populations facing high levels of predation mature more quickly, reach smaller sizes, produce more but smaller offspring, and die more quickly than guppy populations facing low predation levels. Subsequent analyses confirmed that these population differences were genetically determined, implying that guppy populations were locally adapted to the predation regime. The implications for evolutionary change were subsequently explored by taking advantage of natural conditions in the field. Two streams had waterfalls below which there were guppies facing high predation and above which there were no guppies and few if any predators. So guppies were transplanted from below the waterfall to the low-predation environment above the waterfall. After eleven years, the life-history traits of the descendants of the original transplants, like age at first reproduction and offspring production, matched those observed for guppies from other naturally low-predation environments.

Further experiments using individuals from high- and low-predation environments reveal predictable ecosystem level consequences associated with the population level differences. Their different morphologies meant that they selected different kinds, sizes, and amounts of prey. These dietary differences had knock-on effects as well, causing differences in the amounts of nutrients excreted and cycled within the ecosystem. Ultimately there were 20 to 40 percent differences in levels of ecosystem functions like primary and secondary production and decomposition; and in ecosystem properties like algal and invertebrate biomass.

The grandest experiment of the human enterprise is yet to completely unfold. Even with substantial efforts to reduce greenhouse-gas concentrations, many of Earth's ecological systems face substantial future warming. Consequently, there is much concern that the inability of species to tolerate or adapt to changing climate may cause ecological communities to disassemble as species migrate or to go extinct. Such potential loss forebodes subsequent erosion of ecological functioning, which in turn risks jeopardizing many environmental services that support human livelihoods.

On the one hand, this is a pretty grim picture of the future. On the other hand, the realization that evolutionary and ecological processes operate contemporaneously offers some hope that species have the capacity to adapt and thereby sustain ecological functioning. Species populations are often spatially separated along temperature gradients within their geographic ranges and thereby potentially have different thermal tolerances through local adaptation. This arises when organisms within populations become adapted to the range of temperatures they most commonly experience. Evidence shows that the temperature ranges that organisms within populations tolerate tend to match the variation in environmental temperature they experience. Thus, temperature tolerance ranges appear to have evolved locally to be as narrow as possible to minimize the energetic costs of maintenance. Moreover, individuals within species populations may have additional acclimatization potential that allows them to rapidly adjust their physiology, through phenotypically plastic responses, and thus to endure episodic warming events. Again, such plasticity may be an immediate survival mechanism that enables populations to become genetically adapted to chronic and lasting warming.

Such population structure within species means that heat-tolerant populations could replace losses of heat-sensitive populations. At the very least, species presence could very likely be preserved within communities across broad landscapes in a warmed world. The presence of species is maintained because there is a broad portfolio of genetic variety (locally adapted populations) available that can be selected to maintain the species in the face of change.

We have come to learn, however, that populations that have become adapted to local environmental conditions may cause different levels of ecological functioning. In the extreme, they may even have evolved altogether different functional roles. The looming uncertainty is whether in a functional sense "like would replace like." That is, will populations of a species with different thermal tolerances also differ in the nature and strength of their functional roles within communities? Ecologists are now deploying transplant experiments that test for local adaptation and plasticity in thermal tolerance among species populations and whether this adaptive capacity is sufficient to sustain biodiversity, food web dependencies, and ecosystem functions along with services across different thermal conditions.

* * *

Domestication of nature by humans is transforming landscapes. It is pushing nature beyond limits and bounds toward states that have heretofore never been seen. Classic ecological principles stir up much anxiety that the Earth's biota is doomed to rapidly go extinct because of an inability to keep up with the pace of change. Yet, the New Ecology reveals that species may rapidly evolve and adapt to their changing environmental conditions. This gives hope that the future may not be as dire as is often portrayed.

Relying on evolutionary processes as a means of "rescue" has its upside and downside. Genetic diversity, which allows evolutionary processes to happen rapidly, may be key to keeping species resilient and ecosystem functioning sustainable in a changing world. Humanly caused change is highly directional, however, and thus there is a strong predilection to select for subsets among the available variety within species populations. Such selection may winnow variety, leading to genetic homogenization of species populations. Ultimately, this shrinks the portfolio—the variety—of species that is available to make adjustments in times of changing fortunes. It erodes adaptive capacity and leads to loss of sustainability.

At a fundamental level, this means ensuring that species communities do not become relegated to habitat islands, isolated from each other because migration is cut off. Thoughtful planning at the landscape level can configure habitat space and connectivity so that species may persist across landscapes.

CHAPTER 5

SOCIO-ECOLOGICAL SYSTEMS THINKING

Northern Cod stocks were once so plentiful in the waters off eastern Canada and northern New England that when they were discovered during John Cabot's 1497 explorations, it was remarked that they sometimes stopped the progress of ships. It would have been inconceivable at that time that the commercial northern cod fishery would end 500 years later in the spectacular collapse it did. To many, the abruptness of the collapse is all the more surprising because the entire fishing enterprise seemed to be sustainable, and indeed, evermore thriving over most of its 500-year history.

The saga of this fishery offers a historically detailed case study that illustrates how humans and nature can become increasingly entwined as a socio-ecological system. It shows how human exploitation of a resource species can lead to evolutionary changes that alter the productivity of the exploited species, with knock-on effects on the functioning of the ecosystem. It shows how those effects reciprocate back and forth between the ecosystem and the dependent human social system in ways that collapse the fishery and ultimately the human social system. The New Ecology has begun to use historical lessons like this in order to develop better principles about how exploited systems that are entwined with human social systems operate, and to help avert future collapses of exploited systems.

To understand how the changes happened, it is useful to divide up the time course of the northern cod fishery into four major periods of human engagement: the discovery phase (1500–1700); the expansion phase (1700–1770); the post-Revolutionary War nation-building phase (1785–1885); and the commercial- to industrial–fishery phase (1886–1990). It is during this last phase, especially the last thirty years, when factory fishing dealt the final salvo, which ended in abrupt collapse. Each phase marked increasing human dependency on the fishery to support livelihoods and economic well-being. Each phase marked advances in fishing technology and the increasing capitalization of the enterprise. Each phase marked changes in human socio-politics. Together these changes served to increase the yield of this commodity.

The fishery began in earnest between 1501 and 1504 when Portuguese, and then French and Spanish, and finally English fishermen began sailing from Europe to the waters off Newfoundland. These forays typically involved harvests of inshore stocks where fisherman trawled baited long-lines or cast small nets from rowing or sailing dories. Cod were salted and dried on shore all summer long. The fishermen would return to Europe each fall with their holds full of salted fish. The nation that dominated this migratory fishing tradition changed much during this period. European economic conditions, the different wars that were waged, and the peacemaking treaties negotiated between nations determined which fleets would come to fish, and whether any would even come to fish at all.

During the expansion phase, humans gradually built more permanent settlements along the seacoast from New England to Newfoundland. The economies of these communities were originally geared to subsist on inshore dory fisheries. With increased investment, however, the economies quickly grew to become major enterprises by using schooner fleets that were able to compete with the migratory inshore fleets from Europe. Schooners were also outfitted to carry dories to harvest the stocks in offshore waters. Size-selective fishing became instituted during this period, with cod landings graded and economically valued according to size classes. The prized ones, or "great cod," were in the 90- to 100-pound range, the middle-sized ones ranged between 60 and 90 pounds, and the small sizes were less than 60 pounds.

During the post-Revolutionary War phase, humans entered a period of nation building when Canada and the United States redrew national boundaries and began protecting their inshore fish stocks by instituting regulatory measures to control access by foreign fishing fleets. It was a period of urban growth. Rising societal demand meant that more fish were sent to markets in major North American cities and exported to European and Caribbean markets. Economic policies were also instituted to stabilize prices of materials such as salt for preserving harvested cod and to subsidize the fishing communities. Advancements in schooner technology also allowed fish to be salted and packed on board, allowing fisherman to spend weeks to months at sea.

The final phase began with increasing use of capital-intensive technology, including trawling large nets that could scour the ocean floor. The fishery was ultimately industrialized with the advent of the factory ship, which effectively brought an end to the artisanal, inshore dory fishery. Until then, the dory fishery had remained largely unchanged for more than 450 years. These gigantic vessels used emerging sonar technology—a dividend of World War II ingenuity—to electronically pinpoint the locations of the cod stocks. The ships trawled gigantic nets behind them to capture large amounts of fish in a single sweep. The fish were cleaned, filleted, and flash-frozen on board and stored in high capacity freezers. The ships spent months on end at sea and returned fully laden to their North American and European home ports.

Fisheries scientists have accurately reconstructed the time course of the northern cod harvests, using historical records from ships' logs and diaries, and from official government documents. For most of the 200-year discovery phase, the northern cod fishery landed an average 20,000 tonnes (a tonne is about 2,200 lbs) per year, rising up to 50,000 tonnes per year during the last decades of this period. The landings fluctuated slightly from year to year. These fluctuations seemed to be brought about by the varying presence of the fishing fleets from Europe as the vagaries of the socio-political conditions changed there. Cod harvests steadily rose to 200,000 tonnes per year during the seventy-year expansion phase. Year-to-year fluctuations also increased in amplitude. Local inshore fishery landings sometimes were almost nonexistent, indicative of overharvest. Moreover, the average size of cod that were caught began to decline then. The post-Revolutionary phase saw annual harvests grow to over 300,000 tonnes along with greater amplitudes of year-to-year fluctuations. This level of harvest continued for the first forty years of the industrialization phase, yet there were ever-higher year-to-year fluctuations. This was followed by a twenty-five-year downturn reaching a low point of 150,000 tonnes in a single year, then a ten-year upswing when factory ships brought the catch back to 300,000 tonnes per year. Factory fishing led to a ten-year exponential increase in harvests, which peaked at 800,000 tonnes in a single year. This peak was immediately followed by a catastrophic crash to 150,000 tonnes for the year. There was a final upturn reaching 200,000 tonnes per year and then another decline. The entire northern cod fishery halted in the mid 1990s with hard economic and social consequences felt in coastal communities in New England and Eastern Canada, and in Europe.

With the benefit of hindsight, these data offer telltale signs that something was amiss, long before the fateful end. It is a hard-earned lesson. Again, it was human agency pushing things beyond historical trends and limits that led ecologists to uncover new insights.

* * *

The simplest way to understand how the stock sizes of any renewable natural resource (or by analogy with the stockpile of money in a bank account) changes over time is to consider the dynamics in terms of inflows that enhance the stock size and outflows that diminish it. Inflows into cod stocks can come from two sources: immigration of individuals from other stocks (akin to bank account deposits), and offspring production by adults comprising the stock (akin to bank interest). Outflows from the stock are due to emigration of individuals out of the stock and harvest (both akin to bank account withdrawals), and due to natural mortality of individuals within the stock (loosely akin to loss in monetary value).

The financial analogy breaks down, however, when considering stock dynamics of animal or plant populations. Unlike bank savings, where the interest rate does not change with the amount of savings, the production (interest) rate of animal and of plant populations can change as population (stock) size increases. Production rate initially rises when population size increases because more individuals develop to become breeding adults. Population size eventually crosses a maximum, a switch point beyond which production rate steadily declines with further rises in population size. This decline stems from competition among population members for limiting food resources. Competition intensifies with increasing population size because less food is available per individual. The ever-diminishing share of resources lowers each individual's own survival and development, and their production of offspring. This phenomenon is known in population ecology as density-dependence.

A population (stock) with inflows and outflows can be considered the simplest kind of system. Inflows that come from offspring production and outflows that come from natural mortality occur within the system itself. In systems parlance, these are known as feedbacks. Offspring production enhances stock size, so it is considered a positive feedback. Natural mortality due to density-dependence controls the maximum size that a population can reach, so it is considered a negative feedback. Stock size will remain constant whenever enhancing, positive feedbacks are balanced by controlling, negative feedbacks.

When fisheries exploit resources, they are effectively withdrawing from the stock. A human/nature divide is fostered whenever fisheries simply increase or decrease the withdrawal rate in response to changes in human demands or prices, without regard to changes in the harvested population. However, if the fishermen change their exploitation behavior in response to realized returns from harvesting effort, as well as from changes in demand or pricing, they start to become an intertwined socio-ecological system. These principles can help to understand how things began to go awry with the cod fishery.

The first insight is that for most of its history, the fishery perpetuated a human/nature divide. During the discovery phase, the fishery implicitly viewed cod as a renewable commodity, and an everlasting one at that. They harvested cod, but the available technology prevented them from heavy exploitation. Indeed, harvesting may have slightly reduced cod populations (stocks) from their maximum population sizes, in turn causing higher production rates of the stocks. These production rates may have been sufficient to offset the harvest rates, which may explain why there were steady harvests during this 200-year period. Over time, social and economic policies continually favored greater exploitation of the fishery. The increased capitalization of the enterprise demanded profitable returns. If harvest levels fell below targets, humans responded not by backing off to let the fishery rebound but by increasing fishing effort. On top of this, size-selective harvesting changed the size makeup of the adult cohort and hence the productivity of the stocks (see chapter 4).

The second insight is that the fishery responded incorrectly when managing the exploitation of cod stocks. Downturns in harvests, despite fixed fishing effort, are indicative of overharvest, which happens when the production capacity of the stock is insufficient to offset losses due to harvest. The fishery could have responded by setting quotas and thereby imposing a correcting feedback in the social part of the socio-ecological system, to control itself and thereby lower fishery withdrawals (mortality). But quotas were rarely, if ever, imposed. Instead, the technology was advanced to increase the harvest efficiency and the amount captured. In the cod fishery, this new schooner-fishing technological development threatened the livelihoods of the inshore dory fishermen. Schooner technologies stood to reduce the manpower needed to fish, as well as create an economy of scale that could increase the potential catch per effort. A wholesale switch to schooner fishing would have harmed the livelihoods of the inshore dory fishing communities. So policies were instituted to allow limited use of the schooner technology while still freely allowing the large dory fishing labor force to continue. The policies ended up encouraging greater withdrawals. The socio-ecological system was badly misaligned.

Similar policies, and indeed reluctance, to impose strict quotas occurred again in the 1980s and 1990s when the catch-per-unit effort of the heavily capitalized industrial fleets was woefully insufficient to be economically profitable. Imposing any restrictions at that late date in the history of the fishery, however, would probably have been useless. It was likely that the system was already on a trajectory of irreversible ecological change that doomed the fishery to economic extinction. That is, cod are still part of the system—the species is not extinct—but stocks are unlikely to reach levels needed to resurrect an economically viable commercial enterprise.

This leads to the third insight, that this all came about because the fishery used an inadequately oversimplified conception of an ecological system. The idea that a resource population or stock can be treated as a density-dependent system with enhancing and controlling feedbacks is elegant for its simplicity. It was a particularly attractive model for natural resource economics, especially to commercial and industrial fisheries, because it offered a way to operationalize sustainability. According to this view, one uses the density-dependent feedback effect to reach a harvest level—called maximum sustained yield—that maximizes stock production and thereby sustains economic return. According to this model, a previously unharvested population could be carefully harvested to lower its abundance. This density reduction would be compensated by increases in stock production rate. Finely tuned management could be used to lower the population (stock) size down to the point where yield (productivity) was maximized. Then, according to theory, harvesting no more than that productivity would sustain into perpetuity the fish stock, its maximum yield, and human economic return and the well-being that comes with it.

The problem with this view, especially when applied to fisheries, is that it presumes that the harvested population operates as an independent entity. Fisheries typically harvest top predators of food webs. Top predators are dependent on prey species in lower trophic levels for their existence. Cod, as dominant top predators, are no exception. Their immediate prey species are mid-sized predators like hake, squid, crab, herring, and mackerel. These prey species depend on myriad species of zooplankton that in turn depend on phytoplankton at the bottom of the food chain. As added complexity, the small-bodied larval and juvenile cod are fed upon by some of the mid-sized predator species, whose abundances are controlled by adult cod. Thus, adult cod help their offspring reach adulthood by improving their chances of surviving the gauntlet of mid-sized predators.

Harvesting dominant top predators can trigger changes in top-down control of entire ecosystems. Changed top-down control has cascading effects that directly change the abundances of prey species, and indirectly change the abundances of the prey species' resources, and so forth. These changes can lead to countervailing bottom-up feedback through altered flow of nutrients and energy back to the exploited predator population. Thus, harvesting adult cod, especially the larger individuals, could lower the amount of nutrients and energy flowing up the food chain to support cod production. In an interesting twist, it could also mean weakened control over the mid-sized predators that prey on larval and juvenile cod, again diminishing cod population production.

This is but one of many examples in which humans need to view exploited fish species, or any exploited species for that matter, as part of a system of interdependent species. In the context of the cod fishery, loss of large adult cod released mid-sized predators from being controlled, enabling them to exert stronger control over larval and juvenile cod. Such cascading effects in the heavily harvested cod stocks can explain why they seem unable to recover to harvestable levels, despite the moratorium on fishing.

The growing appreciation in ecology that species are part of complex food webs with top-down and bottom-up controls and feedbacks has led to a fourth insight. Moratoria on species exploitation may not lead to recovery. Moratoria are policy instruments that are rooted in the simplified systems model. They are based on the presumed existence of a direct linear relationship between harvest level and impact on the harvested population. Removing the harvest should therefore immediately allow stock sizes to smoothly rebound over time. The longer the moratorium, the greater the recovery.

When species are embedded in food webs, however, the interplay between top-down and bottom-up control and their knock-on effects creates nonlinearities. This means that populations of desired species may recover to harvestable numbers only after they reach some threshold density. It may take a very long time to reach this threshold. Or, the stock may not recover at all because the system has flipped into an alternative state. The fate of cod stocks may be a case of alternative state. Overharvesting, accompanied by the evolutionary changes in adult cod sizes and lifecycles (see chapter 4) may have restructured the food web in such a way that the adult cod are no longer large enough or abundant enough to control the abundance of the mid-sized predators. This particular case of meso-predator release makes it harder for larval and juvenile cod to survive the predation gauntlet, to build up the adult cohort of the population, let alone rebuild the adult cohort size structure. Consequently, the functional role of cod as the dominant top predator in the food web has been supplanted. Recovery of cod stock sizes likely will require first restoring its functionally dominant role as top predator before populations can become highly abundant. Paradoxically, however, cod must become highly abundant before its dominant functional role can be restored. The restructuring of the food web, and the inability of cod to resume its functional, controlling role, is what probably keeps the cod stocks in an alternative state.

Given the ecological, social, and economic upheaval associated with a potentially unrecoverable fishery collapse, it would have been helpful to have some leading indicators that foretell a shift to an alternative state. Until recently, such indicators were not available. The reigning paradigm of maximum sustained yield did not allow even for the possibility of alternative states. The availability of long-term fishery data, like those reconstructed for cod, has provided ecologists with new information to better understand the dynamics of complex systems. That information, combined with data on food web changes that accompany exploitation, has spurred the development of new computational tools that can account for feedbacks and nonlinearities. These tools have led to new principles that can help ecologists anticipate state changes in complex systems and thereby better guide human engagement with nature's economy.

* * *

Ecologists' ability to understand the dynamics of complex ecological systems has been aided tremendously by modern advances in mathematics and high-speed computing. We can now rapidly analyze the multiple mathematical equations needed to characterize the dynamics of species in highly interconnected, complex food webs. The models are used to predict how feedbacks propagate throughout food webs in response to disturbances, such as harvesting.

The important insight is that maintaining a long-term sustainable harvest requires monitoring not only the abundances of the harvested species but also, perhaps counterintuitively, the prey of the harvested species; and furthermore the resource of those prey. Prey species lower down in the food chain, such as zooplankton and phytoplankton, which can reproduce within weeks to months, have faster life cycles—their dynamics fluctuate many times more within years and among years—than annually breeding cod. These smaller species are highly sensitive to disturbances. Problematic disturbances can then be rapidly detected because of highly changeable fluctuations in these species' abundances. They offer valuable, early detection that something is starting to be amiss.

The effects of fluctuations in abundances of species lower down in the food web can eventually flow back up to influence the amplitude of year-to-year fluctuations in the harvested population. As a consequence, the amplitude—volatility—of year-to-year fluctuations in the harvest, not the mean harvest return, is what tells whether the enterprise is sustainable. According to this new principle, important course-correcting warning signals for the cod fishery were, in hindsight, already evident during the 1880s to early 1900s when the upswings and downswings in year-to-year harvest returns continually grew larger.

The challenge for decision making is that there are other factors that may be responsible for causing fluctuations in the exploited system. These include year-to-year variation in climate, and variation in the behavior of humans in socio-ecological systems. This added complexity makes it hard to decide which factor is solely or most fully responsible for driving the system's dynamics. Such uncertainty creates reluctance on the part of policymakers to act.

These added factors increase the complexity of the ecological models. But modern computational technology has enabled ecologists to explore different scenarios of outcomes in which each of the factors is considered alone and in combination with the others. It helps to reduce uncertainty about how much is humanly caused and how much is due to natural processes. Scenario analyses helps us imagine future conditions or outcomes. It can offer guidance on appropriate ways to intervene in a system to achieve desired outcomes. As well, it helps to clarify the potential risks and costs of taking different actions, or even of no action at all. In some cases, inaction can be highly costly. In the case of the northern cod fishery, there are two globally important costs of not acting soon enough to regulate fish harvests.

First, the consequent restructuring of the marine food web has altered the species of phytoplankton and zooplankton that are now dominant. There is especially greater representation of smaller-sized zooplankton. Zooplankton size determines the size of phytoplankton they are able to consume. So, shifting zooplankton size structure has cascading effects on the abundance and size structure of phytoplankton. This in turn changes the fate of carbon in the ocean.

Historically, the phytoplankton community was dominated by diatom species that, in the course of their lifecycles, provided a huge ecosystem service by taking up carbon during photosynthesis and transporting that carbon within their tissue into the deep ocean where it could be stored for long time periods. This way that carbon is sequestered in an ocean ecosystem is known as the biological pump. The restructured food web has now altered this greenhouse gas regulating service by favoring algal species that remain in surface waters. This change in the food web, accompanied by decreases in nutrient upwelling from the deep ocean due to climate-altered thermodynamic properties of the water column, has lowered the supply rate of nutrients to the surface waters. As a consequence, the rate of photosynthetic carbon uptake by algae is becoming limited by low nutrient availability in the surface waters. Ultimately, algae are taking up less carbon than the diatoms did. Moreover, much of the carbon absorbed by the ocean resides longer in the surface water. Together, chemical reactions in the surface water, and decomposition of algal organic matter by bacteria there, make it highly likely that carbon will be released back to the atmosphere rather than being stored in the deep ocean. This change in the biological pump is evident across the entire North Atlantic Ocean region.

Second, to meet the continued demands for fish in European markets, factory-fishing fleets now migrate to another lucrative fishing ground off the coast of Ghana in West Africa. In a case of history repeating itself, these fleets are competing with another inshore, small boat artisanal fishery. In Ghana, fish provide an important source of protein for local communities. Limited availability of dietary protein is already a significant and chronic human health issue there. Protein deficiency specifically causes developmental abnormalities in children with effects lasting well into adulthood. By decreasing fish supply to local markets, industrial-scale harvesting has forced local communities to seek protein elsewhere. The biggest source is bushmeat from Ghana's tropical forest hinterland. Bushmeat comprises all kinds of reptiles, birds, and mammals, the latter including rare deer species and great apes. These species belong to food webs that help maintain functionality of the tropical forest. Human agency promoted by policies in far-removed places has tied the fates of a marine and a tropical forest ecosystem and of the local human populations that depend on both for their livelihoods and well-being.

As ecosystem engineers, humans are averse to uncertainty and the decision paralysis that comes with it. Throughout our history, humans have been inclined to extend classic engineering ideas of command and control, or of single cause/single effect, whenever they intervene in a system. The intent is to make the delivery of a particular ecosystem good or service less uncertain. The repeated lesson, however, is that nature's complexity, and the variability that comes with it, will inevitably confound any efforts at tight control. Such approaches end up doing little more than making ecosystems brittle and thereby prone to collapse. The new appreciation for nature's rambunctiousness requires accepting and embracing variability head-on, scientifically understanding the factors that contribute to determine its magnitude, including human exploitation of nature, and managing within those limits. This new view of the management of complex systems is embodied in the concept of resilience.

* * *

Resilience is the capacity of a system to absorb shocks from disturbances and yet sustain its structure and functioning. The objective of sustainability, in an ecosystem-resilience context, is not to explicitly deliver a steady supply of some single commodity (e.g., fish or lumber). Rather, new thinking holds that the objective is to maintain the functional integrity of the ecosystem. The commodity, then, becomes one of many spin-off dividends of sustaining the functioning of whole ecosystems and the services that they provide.

Ecosystems provide many functions and services simultaneously. There are supporting functions like soil development, nutrient cycling, and energy and material flows. There are regulating services like water and greenhouse gas regulation, including carbon sequestration. There are provisioning services that produce material dividends like fish and timber. All are interdependent.

For example, forestry operations that grow and harvest certain desired tree species would not be sustainable unless there was already production of healthy soils and nutrient cycling to support tree production within the ecosystem. The enterprise would also not be sustainable if greenhouse gas emissions weren't regulated to ensure that the temperature and moisture conditions remain within the evolved tolerance ranges of the desired tree species growing in that ecosystem. It would not be sustainable if wildlife species, as integral parts of the food chain, weren't able to assume their functional roles. Sustaining the several kinds of ecosystem services simultaneously is necessary to keep forest ecosystem productive, and forest operations profitable. Taking such an approach to forest management has the added benefit of perhaps lowering financial operating costs whenever natural ecological processes, especially species interactions in food webs, are enlisted as part of the management activity.

Consider again boreal forests. They are composed of mixes of aspen and spruce, which are economically important natural resources to humans. Aspen is used in plywood manufacturing and in pulp and paper production. Spruce is used for dimension wood: the 2×4s and 2×6s used in housing construction. Successful regeneration of both tree species following harvesting is therefore a key management objective. But for many companies it is a vexing problem because aspen, a competitive dominant species, suppresses regeneration of spruce, often leading to aspen monocultures.

The reason that mixed wood regeneration often fails is that the forest industry may not have developed the appropriate conceptualization of the functional system. The traditional commodity production view focuses merely on growing trees. It involves clear-cut harvesting over many hundreds to thousands of hectares, followed by extensive after-harvest site preparation for replanting using costly heavy machinery, followed by intensive replanting of costly nursery-grown spruce seedlings. The industry also tries to keep large herbivores like deer and moose at bay, by controlling their population sizes. This stems from the belief that, without such control, these herbivores will eat the valuable regeneration. The new, alternative view, however, recognizes and enlists ecological interactions among the herbivore and tree species to foster ecosystem functioning. It treats production of the mixed-wood timber supply merely as one dividend.

In boreal forest ecosystems, deer and moose move around landscapes to forage and to evade large predators like wolves. Both herbivore species prefer to forage on aspen and thereby can potentially suppress its dominance. Experiments that fence out these herbivores reveals that they do indeed mediate aspen/spruce competition. But clear-cut harvesting changes the forest landscape in ways that influence deer and moose foraging behavior. In particular, large openings increase the vulnerability of these herbivores to hunting by both wolves and humans. These herbivore species respond by moving to the edges of clear-cuts to seek refuge within the remaining intact forest. They will forage only at the fringes of the clear-cuts. Aspen often then dominates most of the cut-over area.

The risk perceived by the herbivores can, however, be altered by changing the way that forests are harvested. If landscape were harvested to leave remnant habitat patches that range from two to five hectares in size, throughout the larger cut-over areas, then deer and moose would have many refuges scattered across the entire landscape. Experimentation shows that retaining such patches can draw moose and deer out from the forest edges. Then, through their foraging, they mediate aspen/spruce competition across the entire harvested area, fostering more balanced mixed-wood regeneration.

The changing foraging behavior of deer and moose in response to changing levels of perceived predation risk is another example of phenotypic plasticity. Phenotypic plasticity allows these animals to adjust and sustain their survival and reproduction: they adapt to changing environmental conditions. Their adaptive responses in turn cause knock-on effects that influence the species makeup and functioning of the ecosystem. Thus, focusing management to keep the species interactions intact builds whole ecosystem resilience that in turn sustains the production of the desired commodity.

This example illustrates how ecosystem resilience emerges from an inherent nested hierarchy within the inner workings of the ecosystem. It involves interactions occurring on different time-scales. Individuals within species respond to change or disturbances by exhibiting adaptive responses. Within a very short time, they can change their morphology, behavior, or physiology to maintain survival and reproduction, and thereby their functional roles in ecosystems. The degree and nature of their trait changes determines how these species self-organize, to form and maintain their connections in food webs. It determines how they interact with each other over a growing season or longer.

The legacy of these interactions may carry over to affect nutrient cycling for years to decades. Ultimately, this hierarchy determines the properties and characteristics of an entire ecosystem, say the long-term tree species make up of a boreal or tropical forest, or the nutrient hotspots around termite mounds that create the food web structure and patterning of savanna landscapes. These ecosystems endure for decades to centuries, seemingly unchanged, or changing very slowly, giving the impression that they are in a steady balance of nature. But the apparent steadiness is a product of species being rambunctious, continually in flux, rapidly adapting to ever-present environmental changes, and thereby maintaining sustainability by helping to buffer ecosystems from disturbances. Ecosystems are thus not only complex they are adaptive. In systems-thinking vernacular, they are called complex adaptive systems.

Natural disturbance regimes that create environmental change become important determinants of species' adaptive capacities. Disturbances select for variety—diversity—in the capacity of individuals within species, and of species within ecosystems, to respond to a wide range of conditions and thereby sustain ecosystem functioning. Controls that remove such disturbances, such as wildfire suppression or river channelization to halt episodic flooding, might seem to be effective from a purely commodity production perspective. But such controls are altogether antithetical to sustainability from the standpoint of resilience.

Resilience thinking, however, recognizes that buffering that comes from adaptive capacity of species is not a panacea. The capacity for species to adapt via phenotypic plasticity or rapid evolution is rooted in their genetic makeup. Genetics limit the extent to which any species can change. Change that pushes these capacities beyond their limits, like harvesting that selects individuals with certain traits, or excessive pollution levels that exceed individuals' physiological stress tolerances, can erode these buffering effects. The risk is that ecosystems could then flip into alternative states. A current challenge in ecology is developing deeper, more precise empirical understanding about how limits on the overall scope of ecological functioning is determined by species' evolutionary adaptive capacities, as determined by their genetic makeup and evolutionary history.

* * *

Resilience also demands that we abandon the notion that there is a human/nature divide, where humans somehow exist outside of ecosystems, not as part of them. This idea of human/nature divide derives from an ecological paradigm that excludes humans as part of the system. It also came from a scientific experimental tradition in which a focal "natural" system—say a food chain comprised of a diversity of insect predators and herbivores and herbaceous plants—is isolated. The focal system was then probed and pushed to test how disturbances, like removing top predators or increasing soil nitrogen supply within the system, drive the system's dynamics. Once this was understood, the effects of other, perhaps humanly caused, disturbances like pollution or herbicide application could be tested. But human-caused disturbances were viewed as coming from outside of the system. This is what has led to the notion that humans habitually disturb ecosystems.

This "divide" is further promoted by classic economic thinking in which the ecosystem is viewed as being external to human society. Classical economics focuses on societal interdependencies and interactions through the economic market. Ecosystems are then viewed as a source of external inputs of resources into the market. They offer provisioning services that provide the raw materials that influence financial use values. This view presumes that changes in the use values for those materials provide the self-regulating feedbacks that prevent society from over-exploiting the resource and causing ecosystem damages.

Both perspectives ignore the fact that humans too are biotic species and, as such, have many evolved tendencies. Humans follow game plans, as do other species. They have the capacity for extremely selfish behavior as well as for selfless and cooperative behavior. These behaviors are adaptive. Perhaps not in the sense of the tight evolutionary game that nonhuman species must play to survive within nature's economy, but such adaptive behaviors shape values and attitudes, norms for social behavior, and policies that determine how societies operate. The inner workings of human social systems are also driven by nested hierarchies that shape society's adaptive capacity and therefore its resilience to disturbances within the social system itself, like volatility in the market, as well as disturbances from without, like storms that cause devastating property damage.

But societies still tend to hold to the world view that humanly based and nature-based systems are largely independent entities. These world views ignore the fact that human behavior can cause reciprocating feedbacks between the market economy and nature's economy. Socio-ecological systems thinking instead recognizes that such human behavior causes society and nature to be entwined in fascinatingly complex, adaptive systems. Thus market uncertainty and volatility can in part be tied to volatility in commodity production, which is tied to the way that humans domesticate and engineer nature.

For example, mining in Bristol Bay could provide sorely needed employment as well as a huge supply of metals that are critical to the global technology market. But it risks polluting the headwaters of rivers that support spawning salmon, and thereby risks collapsing the annual salmon runs there. Such loss would have important feedbacks for the social and ecological systems in the region and beyond. Loss of salmon runs could lead to the loss of important nutrient inputs that support up to 150 different animals species that are organized into food webs within the river and surrounding forest ecosystems. It risks the loss of employment in communities whose livelihoods and well-being depend on sustainable commercial and recreational fisheries in the region. It risks losing the nutrient subsidies that flow from the river to the surrounding forests. This, in turn, risks lowering forest production that supports employment in a valuable forest industry there, let alone that it supports the supply and price of timber commodities that enter the global market. Global demands for re-

sources from one region can impact global supply of other resources from that same region. The livelihoods and well-being of humans in that region, as well as the other global regions that are linked through exports and trade, hang in the balance.

Barry Commoner, one of the early thought leaders in the environmental movement, is widely known for proclaiming, in his book *The Closing Circle*, that an indubitable law of ecology is that everything is connected to everything else. Many may view this as rather glib. But a new socio-ecological systems thinking inescapably embraces and builds on this reality. It is part of the fundamental principle to understand emerging, reciprocal feedbacks that ultimately determine the resilience of human social and natural systems and hence global sustainability. Appreciating reciprocal feedbacks also means overcoming the tendency simply to equate "social" with economics and the inherent financial values, choices, and efficiencies that drive human behavior. *Social* embodies many more aspects of humanity than just economics.

Human social systems are an amalgam of natural and social factors that determine the capacity for adaptation. Social systems depend on natural resources (energy, wood, water, and so on), socio-economic resources (labor, capital, information, and its communication), and cultural resources (myths and beliefs, religion and ethics). Together they create hierarchical social structures that steer human behavior. At the base of the hierarchy is basic individual neuro-physiology that determines individual human behavioral cycles, human health, human values and preferences, and human actions. At another level, there are social institutions that address needs for health, justice, faith, and sustenance. Together, the individual and institutional parts of the social system create social order. This order emerges from changeable interactions among people and groups and how they shape the norms, ethics, and regulations that determine how society functions.

Humans are tied to nonhuman beings in nature's economy by the extent to which societal functioning is dependent upon natural resources and ecosystem services, which is dependent upon the levels of ecosystem functions, which is dependent upon biodiversity. If human social systems and natural systems were weakly linked, then the two, in a functional sense, may be viewed as independent entities. It would be fine to consider the world as a human/nature divide in which humans live within the human "ecosystem" and all other organisms in natural ecosystems. But, even in nature, species within one ecosystem are often inextricably tied to species in other ecosystems through cross-boundary flows of materials and energy. An increasingly domesticated nature has increased those flows and interdependencies between natural and human social systems. These interdependencies can stretch across the entire globe.

Some may despair that all this integration and complexity means that things are getting so complicated that it is increasingly likely that we will never be able to guide human behavior to achieve a resilient, sustainable future. Instead, they try to influence society simply by highlighting the destruction that is being wrought and shaming humans into changing their ways, or admonishing them if they don't. This merely serves to promote a no-longer-tenable human/nature divide.

The New Ecology offers something else and something more. Highly complex is not synonymous with highly complicated. Complexity merely means that simple command-control, single-cause/single-effect views of policy development and environmental management won't work. Complexity helps appreciate the need to privilege the entire system over any single part of it. Complexity helps appreciate the feedbacks that create nonlinearites in system responses to disturbances. The New Ecology shows that understanding the workings of the whole system can be understood and be predicted by understanding the nested hierarchy and mechanisms of the inner workings of socio-ecological systems and the nonlinear responses that arise when they are disturbed or exploited. Ecologists continue to help develop new technology, new tools, and new principles to help society become more thoughtful and effective in their engagement with the environment. Rather than promulgate inaction, ecologists are increasingly helping humans become thoughtful stewards of their environment.

CHAPTER 6

HUBRIS TO HUMILITY

An important way that I scientifically uncover and explain nature's mysteries is through experiments that test my intuitions, which derive from my natural history observations of species in action. More scientifically said, the intuition is reformulated into new theories that produce logical, testable predictions that then guide the design and execution of an experiment that lets me see whether the idea has any merit. If it does, then I have learned something; if it doesn't, then I also have learned something: that I was wrong, that I must go back to the drawing board and think things afresh. This is the way that the scientific knowledge process advances our understanding of how nature, in all of its complexity, works.

The process of discovery through such trial-and-error testing isn't fundamentally different from what many of us experienced during our childhood days in elementary school, doing science fair projects. Indeed, one of my fondest memories of the thrill of discovery through experimentation was when I learned back then what it takes to build a self-sustaining ecosystem. It sealed my fate to become an ecologist.

The experiment involved assembling various ecosystems in glass containers by combining water, with nutrients such as nitrogen, some bacteria, some aquatic plants, and some animals such as snails or insects, after which the containers were hermetically sealed. These glass-contained ecosystems were economies involving production and consumption of food energy and nutrients, and balancing of gas concentrations. In this economy, plants would take up carbon dioxide during photosynthesis to produce edible tissue and oxygen, and they would respire some carbon dioxide in the process; animals would eat plants and other animals and respire carbon dioxide; old individuals would die and the chemical constituents of their body would be broken down by bacteria and recycled through the system. Bacteria also respired carbon dioxide. All the respired carbon dioxide fed back into plant photosynthesis and new production. This cyclical economy operated solely within the fixed bounds of the glass container. And, once the containers were placed in sunlight and temperature that was tolerable to the organisms within, the ecosystems would magically become self-sustaining, perpetuating the cycles over and over again for quite some time.

This same kind of science experiment was tried in the early 1990s, albeit on a very much larger scale, at a fully enclosed glass facility at the foot of the Santa

Catalina Mountains, about a half hour's drive from Tucson, Arizona. Within an area of two and one half football fields, this facility, known as Biosphere 2 (so-named because Earth is Biosphere 1), was used to see whether human ingenuity and technological knowhow was up to the task of replicating nature by sustaining the functioning of several ecosystems all at once, solely within the giant "glass jar."

It was an awesome feat of engineering. The enclosed space contained several miniature ecosystems comprised of plants, and of invertebrate and vertebrate animals belonging to a rainforest, an ocean and accompanying coral reef, a mangrove wetland, a savanna grassland, and a fog desert. The heating and cooling systems relied on passive solar and on-site natural gas generation. Most importantly, space was deliberately allocated for human habitation and agriculture. A crew of eight adventurous people, including a medical doctor and researcher, was sealed inside the facility for a period of two years, during which time they were pretty much left to go about their daily lives. Scientists monitored their health as well as the air, the water, and the surrounding soil chemistry.

The human inhabitants grew bananas, papayas, sweet potatoes, beets, peanuts, cowpeas, beans, rice, and wheat, which comprised approximately 83 percent of their total diet. The diet was supplemented with meat from goats, pigs, fowl, and fish. Monitoring showed that the inhabitants' health improved after entering the facility, including a lowering of blood cholesterol and of blood pressure, and their immune function was enhanced. However, almost from the beginning, the inhabitants complained of hunger. During their first year, they lost an average 16 percent of their initial body weight. They regained some weight during the second year as their metabolism became more efficient at extracting nutrients from the low-calorie, high-nutrient diet as well as by improving their food production capacity. (This is a nice example of adaptive physiological and behavioral plasticity in humans.)

The ecosystems, however, became underdeveloped or transformed because of the internal biophysical environmental conditions. The fog desert became more chaparral-like due to condensation from the physical structure of the facility. Rainforest trees grew rapidly. But they and the savannah trees ended up having flimsy structure due to low sunlight conditions and a lack of mechanical stress from disturbances like wind that normally causes trees to adapt and strengthen their physical structure. The coral reef thrived. But it continually had to be tended in order to remove choking algae and to balance calcium carbonate and pH levels to prevent ocean acidification. The mangrove ecosystem also developed quickly. But it had less understory vegetation than normal, possibly because of reduced sunlight levels.

The most significant challenge was maintaining balanced levels of atmospheric carbon dioxide and oxygen. Carbon dioxide levels fluctuated wildly. Wintertime levels were almost ten times higher than what

we experience on Earth today, and summertime levels were almost triple that of twenty-first-century levels. Oxygen concentration steadily declined to levels found in the thin air at the highest mountain elevations, which is dangerously low for humans. The human inhabitants tried to manage atmospheric carbon dioxide levels using technology (a CO_2 scrubber). They also tried to promote the gas regulation service of the desert and savannah ecosystems by transforming them to cultivate fast-growing plant species to increase photosynthesis. The intention was to enhance carbon dioxide uptake and oxygen release. They also cut and stored the rapidly growing plant biomass to sequester the carbon. But the lack of space to store and protect the ever-accumulating harvested biomass would eventually make it vulnerable to microbial decomposition, with attendant release of carbon dioxide back to the atmosphere.

Further scientific investigations showed that carbon dioxide was reacting with exposed structural concrete inside the facility, leading to carbon and oxygen sequestration as calcium carbonate. This further compromised the ability to restore the balance of atmospheric carbon dioxide and oxygen.

The human inhabitants started to experience sleep apnea and chronic fatigue. Many of the bird and mammal species and all of the pollinating insects died, and this was attributed to the changed atmospheric conditions. Ant and cockroach numbers exploded and morning glory plants overgrew the rainforest ecosystem, choking out other plants. The experiment was halted after two years because Biosphere 2 no longer provided safe living conditions for the humans.

This was a costly venture. More than US $200 million was spent to build it and try to support the health and well-being of what would not even be considered a small village of humans residing within a small acreage of land. It would be mind-boggling to extrapolate what it would cost to artificially support the rest of humanity this way across the entire Earth. Furthermore, given the way of life within Biosphere 2, such investment would support only the very basic necessities of life.

Biosphere 2 provides an important lesson about the limits of technology, and about human ingenuity not always being able to create technology immediately when needed in order to solve environmental problems. Biosphere 2 taught us all too well that we are still far from being able to engineer a functional natural economy that supplies a sustainable set of services to support the health and well-being of all of humankind, let alone all of nonhuman life.

Fortunately, we still have Biosphere 1. Through ecological and evolutionary processes, it already has the diversity of species and associated ingenuity to maintain the biophysical conditions to support human livelihoods and well-being. It provides resilience needed to adapt to change. The trick is to ensure that all of this capacity is maintained as society increasingly domesticates nature. The New Ecology is making some inroads with another form of socio-ecological integration. It is doing this by following in Aldo Leopold's footsteps and enhancing a scientifically informed ethical awareness of human engagement called environmental stewardship.

As discussed earlier in the book, the ethical awareness and non-economic values that humans have for nature plays an important part in shaping human attitudes and behavior: how humanity views and treats life on Earth. Any discussions or debates about whether this or that species, their functional roles, and the services they provide, be conserved must first clarify the ethical viewpoints of the participants. Do people take a strictly anthropocentric (human-centered) view, where everything exists merely to support humankind's utilitarian needs, where nature is just a collection of species for humanity's passing enjoyment? Or, do we take a wholly nonanthropocentric view where humans are considered part of something bigger and thus must respect all of the other parts? Or is it something between these two views? Understanding what environmental stewardship hopes to accomplish requires putting it into the context of these broader anthropocentric and nonanthropocentric ethical considerations.

* * *

By and large, anthropocentric ethical views toward nature and the environment originated and developed from religious contexts where belief in divine command guided the use and protection of nature; or from more secular contexts where guidance was based on rational thought about human engagement with human biophysical surroundings. In both these views, however, ethical standing was largely accorded only to humans, or perhaps extended just to all sentient but still rational beings. Given scientific understanding and even belief systems at that time (e.g., the ethics of vivisectionist philosophy effectively proclaiming that animals are automatons that feel no pain), this ethic excluded almost all animal species; and most certainly excluded all plants.

The field of nonanthropocentric environmental ethics emerged in response to a desire for greater humility in human engagement with nature. Nonanthropocentric environmental ethics develops guidance for human action beyond simply the betterment of human well-being through dominion over the natural world. From a practical standpoint, there are several ways that nonanthropocentric ethics are expressed in society.

Perhaps the most familiar comes from the animal rights and animal welfare movement. An animal rights ethic extends ethical standing to all animals that are believed to be sentient. Typically this extends standing only to reptiles, amphibians, fish, birds, and mammals. It is perhaps no coincidence that these groups of species share an important feature with humans: they all have backbones. These include species—like bears and wolves and wolverines, or lions and tigers and leopards, or zebra and rhino and wildebeest and elephants, or pandas, or whales—that most captivate our fascination with nature. The potential harm to or loss of these charismatic animals is what motivates much of the action to conserve and protect nature. But even collectively, the 80,500 species of reptiles, amphibians, fish, birds, and mammals represent only a tiny fraction of the approximately 8.7 million species living on Earth. Such a view, especially, does not even begin to consider extending deliberate ethical standing to plant species that form the habitats of these vertebrate animals. An animal rights ethic privileges a few species over many others.

An alternative biocentric (or biophilic—meaning *love of biota*) ethic gives standing to all living beings, especially ones that are normally not thought of as having sentience, and, for most humans, certainly not having charisma, e.g., soil bugs, microbes, and fungi. A biocentric ethic holds that all living things have inherent value that comes from being alive and striving to stay alive. This leads to the ethical attitude that all living beings, humans and non-humans, have equal worth. And so, humans should show humility and respect toward all beings because each has a "good of its own." Such an ethic promotes respect for all of nature's biodiversity. Such an ethic recognizes that living beings can be benefited or harmed by human agency. Humans should therefore act in ways that do not deny species their potential to live and thrive according to their biological natures, to assume their natural tendencies.

An ecocentric ethic takes things a step further by extending ethical standing to all living beings and to the nonliving physical components (e.g., geology, soil, water) that support life. Here, respect and humility toward nature includes an appreciation that all living beings are interdependent. Humans and nonhuman species build up the physical and biotic components of ecosystems through their functional roles and interdependencies in food webs. Sustaining these interdependencies is considered essential for the existence of all living beings. They are sustained by allowing all species to fulfill their biological functions.

But for some, this creates a dilemma. If certain acts—say, predation or infection with disease—cause harm to others, should these be considered as part of system function? Should we protect this? An ecocentric ethic would say "Yes," given that a fundamental law of ecology, if there is such a thing, is that every living being has to eat something in order to exist (in ecology disease is considered merely a variation of predation in which the victim does not necessarily, die). An ecocentric ethic thus recognizes that if humans are to be considered part of nature, they, like all other species, should have the right to exploit it.

An ecocentric ethic is the basis for modern ecological systems thinking, that humans as biotic species are functional parts of complex adaptive ecosystems. While recognizing humanity's right to exploit nature, such an ethic is not intended to give humanity license to exploit ecosystems without regard to sustainability. Systems thinking teaches us that to maintain sustainability of the whole system, humans must act in ways that preserve food web structure, and also preserve the dynamism created by species interactions and feedbacks. This gives rise to a modernized version of Aldo Leopold's general ethical principle—his golden rule, if you will—intended to guide human action toward nature's economy:

> A thing is right when it tends to preserve the integrity, stability [resilience and sustainability], and beauty of the biotic community. It is wrong when it tends otherwise.

Socio-political and economic interests often motivate human actions. Humans can and do use technology to override species interactions and feedbacks when exploiting resources. Ecologists increasingly recognize that such factors that are outside the sphere of nature's economy can jeopardize its integrity, its resilience and sustainability, and its beauty. In response, ecologists increasingly appreciate that they cannot go it alone in the cause for sustainability. Whether people embrace scientific guidance, or welcome or trust it, depends on social contexts and a diversity of ethical positions held within society. In appreciation of this, ecologists have helped develop a new ethic of environmental stewardship that respects those various ethical positions and the nature of the social systems that determine them.

* * *

Environmental stewardship is an emerging ethic that is intermediate between anthropocentrism on the one hand, and ecocentrism on the other. In a sense, it blends a nonanthropocentric ecocentric view with an anthropocentric one by proposing that humans have ethical obligations to one another that are mediated through their mutual relationships with the environment. Thus, humans have an ethical duty to other humans to protect the integrity, resilience, sustainability, and beauty of nature. Accomplishing this means that, necessarily, they have an ethical duty to protect the interdependence of living beings and the physical environment that sustain natural economies.

Environmental stewardship is not just another name for conservation, the primary goal of which is simply to preserve and protect the species and habitats that make up ecosystems. It is also not just another name for management, the primary goal of which is to minimize potential damages created by society as it exploits ecosystems. An environmental stewardship ethic strives for continuous improvement of environmental performance for the production of ecosystem services. This is accomplished through a commitment to use natural resources wisely and efficiently as dividends of ecosystems functions, to protect entire ecosystems not just their parts, and to ensure development of sensible environmental policies and regulations in support of ecosystem service provisioning.

Stewardship is about finding creative and scientifically defensible actions that minimize risks and maximize opportunities to sustain and restore natural ecosystems and the services they provide for current and future generations. Stewardship involves more than having governments institute laws and procedures for regulating human behavior, and having citizens merely advocating for competing interests. Stewardship encourages all citizens to share responsibility for environmental decision making and to be accountable for the consequences of any actions that arise from those decisions

Stewardship succeeds when people as individuals, or as members of organizations, communities, or other institutions that comprise human social systems, have opportunity to engage with each other in the interest of having a stake in and cooperatively managing ecosystems for long-term sustainability. Sustainability outcomes then emerge from the choices made within a particular social system. These choices depend on environmental, social, and economic contexts. For example, people in poorer societies or in places with meager agricultural productivity, have no options for the use of that agricultural production but to feed themselves with it; whereas individuals in wealthier societies, or in places with highly fertile soils, may opt to dedicate surplus food production to different ends (e.g., biofuels vs. food export). People in poorer societies may also put a premium on feeding themselves over striving to improve water quality.

Ideally, any chosen option should not be permanently locked in. There should be opportunity for societies to make new choices when they see their future fates change. This requires managing for changeability, rather than for constancy, to build the capacity to be resilient, to creatively adapt as environmental, social, and economic conditions change, and to accommodate new scientific understanding. Stewardship operationalizes the idea that humans are part of a complex yet adaptive socio-ecological system.

Effective stewardship requires that citizens take interest in it and take responsibility to stay informed about how their chosen actions may affect their local environment and beyond. To see themselves as being part of a bigger picture. It means appreciating that actions like, say, protecting the environment in a place like Vermont, with its breathtaking, verdant mountain vistas, has consequences far beyond the northeastern geography of the United States in which the state is situated. Some of the mountaintops in Vermont are ideally suited[1] and desirable for renewable energy generation using modern wind generation technology. But windmills are considered by some to be unattractive, gangly pieces of technology. An array of them organized into a wind farm is regarded as an unsightly blight on the landscape. Furthermore, building wind farms requires destroying some of the forested area, which risks losing biodiversity

1 http://www.windpoweringamerica.gov/windmaps/.

associated with the forest habitats. Socially, it can also mean losing a sense of place and community. It is therefore conceivable that some local land trusts and conservation organizations might act to prevent building such wind farms, using arguments that it would jeopardize the integrity and beauty of the forest ecosystem there. It is often argued that building wind farms is inconsistent with the idea of sustainable forestry.

But consider a consequence of such action. Mountaintops in West Virginia and Kentucky are also known for their forested vistas. These forests cover vast seams of coal. This coal is used to generate electricity. In the interest of expediency, coal companies have completely removed many of these forested mountaintops. This has caused innumerable damages to the local environment. Biodiversity and ecosystem functions and services have been heavily impacted there. People have lost their sense of place and community there. Poorer environmental quality has affected human health there. Greenhouse gas emissions to the atmosphere from coal-fired electricity generation affects things everywhere, including Vermont. Is this sustainable forestry? Is it right or wrong that environmental decisions made in one location harms the health and welfare of humans elsewhere through destruction of their ecosystems, their natural economies, and their environmental quality?

An increasingly domesticated world means that societies and ecosystems are less and less spatially isolated units. They are connected across local and regional landscapes and across vast global distances (recall again industrial fishing off of East Africa and the bushmeat trade). They are connected through feedbacks arising from local decisions about land development, exploitation, and trade and how they affect species interdependencies, and the flow of materials and nutrients. The refrain "Think globally—act locally" has never been more apt.

The expanded use of global space as humans increasingly domesticate the world inevitably leads to conflict over what exactly that space should be used for. It can even lead to counterproductive conflict by different organizations seeking similar ends. Consider again the implementation of new renewable energy technology to replace older, fossil-fuel-burning technology. The ideal landscape locations for generating energy from wind, or solar for that matter, often fall squarely within the same landscape locations that harbor some of the most valued or sensitive habitats, or some of the most rare and threatened species. Media stories repeatedly portray conflicts over such locations as David and Goliath-type encounters in which the interests of a small local conservation group or land trust that is trying to protect its cherished species or special place is pitted against a corporate giant with a sole interest in financially lucrative renewable energy generation. These interactions often end up in drawn-out litigation, which encourages prolonged use of fossil-fuel energy with prolonged releases of CO_2 to the atmosphere. Paradoxically, taking actions to prevent siting the newer, cleaner energy-generating sources to prevent species or habitats of concern from being destroyed can increase the long-term risk that climate change will destroy those very species or habitats. And the sustainability of ecosystem functions and services in those locations may go with it.

In the past, the hesitancy to build partnerships between conservation and industry in the interest of common cause or purpose has been attributed, in part, to the absence of critical science. But the New Ecology can offer the needed science and modern spatial analysis tools to support landscape-scale planning for compatible land uses. Ecologists can help craft ways to implement technology alongside maintaining species interdependencies and ecosystem functioning. This is done using scenario analyses, an approach for synthesizing diverse information and describing and evaluating alternative socio-ecological futures. Scenario analyses helps overcome decision paralysis due to uncertainty. It is not prescriptive however; it is not intended to tell people what to do. Rather, it helps decision makers imagine and compare how actions from different choices might play themselves out in the future. It helps to understand the pros and cons—the benefits, drawbacks, and uncertainties—associated with different choices. It can help decision makers understand whether a choice forecloses future options or maintains flexibility and an ability to reverse or to change course as the state of the environment changes. This can help decision makers winnow down decision options to reach a final decision that is scientifically defensible and respectful of societal values.

* * *

Environmental stewardship holds that things must go both ways. Societal values must respect the limits of Earth in order to ensure that the choices lead to outcomes that promote sustainability. This means fundamentally appreciating that we live on a finite planet. The space on the planet must be shared with other species if humans wish to feasibly retain levels of the many ecosystem services that are essential to sustain modern livelihoods and well-being. By advancing a stewardship ethic, ecologists aim to instill an appreciation that humans must live within, not exceed, the means provided by nature. Modern ecological science is devoted to quantifying what those means are in order to identify a safe operating space in which to make choices with respect to use of landscapes and ecosystem functions and services.

The concept of safe operating space supports a commitment to continuously improve environmental performance for the production of ecosystem services. It identifies boundaries or limits based on levels of a suite of environmental factors that are essential to maintain ecological performance and sustainability. These include biodiversity, atmospheric greenhouse gas concentrations, ocean acidity, nitrogen and phosphorus inputs for agricultural production, land use change, and freshwater. This ensures that humans continue to enjoy a portfolio of options for maintaining resilient ecosystem functions. This is why ecologists worry, for example, about the fact that society has already exceeded a safe operating boundary of 350 parts per million concentration of atmospheric carbon dioxide. Exceeding the boundary could start to foreclose options and opportunities for current and future generations, such as sea level rise leaving less and less living space on land, or droughts reducing available arable land, or damaging storms requiring costly repairs of infrastructure.

The concept of safe operating space also highlights that the factors essential to environmental performance cannot be considered in isolation. On a finite planet, change in one variable causes changes in others: they are interconnected and interactive.

A case in point is the aftermath of the cod fishery collapse, where loss of marine biodiversity now risks increasing ocean acidification. This happens because alteration of the marine food web along with changes in the biological pump and nutrient upwelling alter the fate of carbon entering the ocean. As a consequence, carbon is retained in surface waters rather than transported down into the deep ocean to be stored. Chemical reactions with carbon in turn make the surface waters more acidic.

As another example, a major portion of grassland ecosystems in the western United States has been appropriated by humans for cattle grazing and cereal crop production, that latter being heavily reliant on water for irrigation. Historically, large predators like wolves and grizzly bears roamed widely and freely across this landscape. But they preyed on cattle. This threat to human livelihoods prompted wholesale extirpation of these large predators from much of their historical range. The loss of predators is linked to long-term increases in the abundances of native herbivores such as elk and moose. At times, they have increased enough to damage habitat adjacent to rivers through overbrowsing. The loss of species that are reliant on riverine habitat, including beavers, notwithstanding, overbrowsing can alter the physical properties of rivers and streams themselves. Loss of vegetation destabilizes river and stream banks. This leads to bank erosion that lowers water quality. It alters river and stream channels thereby lowering the water table, and the intensity and volume of water flow. This is the very water that is often needed to irrigate cereal crops. Moreover, bringing back the predators may not reverse the damage and restore river ecosystem functioning. This is because restoration of the water table and water flow requires beavers to dam the rivers. However, bringing back the beaver may not be enough. Overbrowsing has led to poor production of woody plants, leading to an insufficient supply of the very material that beavers need to build their dams. Eliminating top predators can cause the river systems to lose resilience and become locked into an alternative state. Eliminating an important predator species to mitigate threats in one agricultural economy risks threatening the sustainability of another agricultural economy by foreclosing its options.

The variables determining safe operating space can be linked across even grander spatial scales. The development of industrial processes to manufacture synthetic nitrogen fertilizer has revolutionized agriculture. Farm policies and financial incentives, abetted by ample supplies of cheap nitrogen fertilizer, have allowed cereal and grain agriculture to expand many times over in places like the southwestern United States. In the forty years between 1950 and 1990, there has been a fivefold increase in soybean yield, and a doubling of corn yield. It has also led to a fivefold expansion of land use dedicated to agricultural production.

This enterprise happens squarely within the wintering grounds of snow geese that annually migrate there from their summer breeding grounds in high arctic coastal areas. Snow goose populations have benefitted tremendously. The added nutrition from stubble, and rice, corn, and wheat kernels remaining after harvest has subsidized their populations and enhanced their winter survival. Snow goose numbers have increased by fourfold to sixfold during this forty-year period. But this has meant that goose abundances have begun to exceed the capacity of the summer grounds to support them. They now overgraze large swaths of arctic sedge and grasslands, leading to loss in plant diversity. With little else to eat, they have grubbed down into the permafrost soil to eat the plant roots. The consequent thawing of permafrost has altered soil conditions, microbial functioning, and nutrient cycling. The extent of the impact is so large that it can be seen in satellite images taken from outer space.

Here, policies and decisions that encourage land use change and alteration of the nitrogen cycling in one location have impacts on ecological functioning in other, remote locations. The fate of the environment for humans across vast distances is, in this case, connected by their impacts on a shared migratory animal species.

The fates of local socio-ecological systems can be tied across vast distances by other activities. Industrial fishing fleets sent from Europe cause bushmeat hunting in Africa. Desertification in northern Africa causes several hundred million tons of dust to be transported annually across the Atlantic Ocean. It settles in the Caribbean, leading to congestion of coral reefs, human respiratory ailments, and loss of soil fertility there. Global trade arising from demand for raw materials or manufactured goods in the receiving location impacts environmental quality and sustainability in source locations where they are mined or manufactured. Ecologists call such long-range connections telecoupling.

The concept of telecoupling effectively extends the idea of food web interdependencies to vast spatial scales. It is a form of scenario analysis that helps operationalize thinking globally when acting locally. As the examples show, it can help to trace the global pathways and potential impacts of decisions that happen within local socio-ecological contexts. It can help to pinpoint the causes of environmental change. It clarifies decision making by helping us understand whether different choices in one location are likely to harm ecosystems in other places and the distant societies that depended on them there, ergo the stewardship ethic.

Telecoupling reminds us that everything is indeed connected to everything else. These are pathways of commerce and trade across ecosystem boundaries. They enhance the dynamism, the rambunctiousness, of ecological and social systems. They are also pathways that propagate feedbacks. For example, trade in surplus food from agriculture may link different regions globally. The source region improves the health and well-being of societies in the receiving system, especially so when the source region is highly productive and the receiving region has meager agricultural productivity. But people in the source region may suddenly choose to reallocate most of their surplus production from export to domestic biofuel production, in response to desires for greater security in energy supply. This is turn feeds back on the receiving society, causing that society to reallocate local land use and water consumption as people clamor to feed themselves. Because of finite space on the planet, actions to increase energy security can precipitate decreases in food security within the socio-ecological systems in the connected regions. An environmental stewardship ethic respects the importance of telecooupling in determining the fates of societies around the globe through their engagement with nature. It encourages reasoning through how decisions in one region may impact the fates of socio-ecological systems in other regions.

* * *

Part of the despair alluded to in the previous chapter, of ever attaining a sustainable future, stems from the worry that human engagement with nature will simply degrade ecosystems. With finite space on Earth, this means there will be less opportunity to walk away from degraded areas and to shift exploitation to non-degraded ones. But the New Ecology is also making important advances through the development of principles and methods to restore and rehabilitate nature. Restoration operationalizes Leopold's medical metaphor for an integrated environmental science and practice. It provides the scientific knowhow and means to repair degraded ecosystems back to their original functioning. Restoration enlists natural processes or develops management to speed up the processes by drawing on scientific understanding of the way ecosystems assemble themselves over time.

Ecosystems throughout the globe originated from natural development processes of primary succession or natural restorative processes of secondary succession. Primary succession follows when species become established on barren substrates like lava flows or glacial remains and then build up to form a natural economy. Secondary succession arises on substrates previously occupied by species after major disturbances like fires and floods denuded the areas. Widespread evidence of ecological succession shows the power of natural processes to re-create ecosystems without help. Ecological restoration harnesses this natural capacity by introducing management interventions that reverse the effects of long-term problems and steer ecological systems back to their original, natural state. The promise of restoration ecology is that it can offer a tool kit of management options to balance environmental protection with providing environmental services for a burgeoning human population.

A reigning sentiment, however, is that it would take ecosystems hundreds to thousands of years to recover, if they recover at all. Certainly, resilience raises the specter that humans can shift ecosystems into alternative, undesirable ecosystem states. And, indeed, I have offered several examples of such state changes. But new scientific evidence shows that this is not the most likely outcome when humans heavily exploit or damage ecosystems.

The evidence is gathered from over 400 peer-reviewed studies that have monitored all variety of natural or managed recovery of ecosystems, after humanly caused disturbances, and they compared how ecosystems return to undisturbed baselines. Whether the disturbance is from agriculture, deforestation, nutrient pollution of water bodies, invasive species, logging, mining, oil spills, overfishing, trawling, or interactions among several disturbances, recovery in most cases can happen during a human lifetime. This holds for ecosystems in tropical and temperate regions worldwide, including freshwater lakes, rivers, and wetlands, estuaries, the marine realm, and forests and grasslands. Only a small set of those studies hint that the ecosystem has flipped into an unrecoverable state. Perhaps there is widespread evidence for recovery because humans have not yet pushed most of these ecosystems beyond safe operating boundaries. An important goal of ecology is to come up with better, more accurate, measures of those boundaries.

In the meantime, humankind will continue to actively domesticate nature to meet its needs. Restoration ecology does not give license to exploit ecosystems without regard to sustainability. But with even the best sustainable practices unforeseen outcomes and damages can, and are likely, to happen accidentally. The message of restoration ecology is that ecologists are equipped to repair the damages. Recovery is possible and can be rapid for many ecosystems, giving much opportunity for humankind to transition toward greater humility in their engagement with nature.

CHAPTER 7

ECOLOGIES BY HUMANS FOR HUMANS

Whenever we push the ON-button of an electronic device there is an expectation that the unit will power up quickly and display images in vibrant colors. If the electronic device is used for communications such as texting, emailing, or web browsing and downloading, there is a further expectation that there will be an immediate response. All this speed and functionality is the result of modern technological advances in the processing capacity of electronic devices, which doubles about every two years (Moore's Law). These advances have created an at times dizzying race by corporations to offer ever-newer consumer products that are faster, have more applications, have a broader network reach, and more.

This capacity increase has come about by wider use of mineral elements. When computer chips were originally developed in the 1980s, they used eleven major elements of the periodic table. Modern computer chips now use about sixty, which amounts to two-thirds of the periodic table! The invention of widely used ear-bud headphones has partly been possible because magnets within them can be made from a rare earth called neodymium. This element helps to produce lots of sound from a very small unit. The use of neodymium, and of lanthanum and dysprosium and other rare earth elements, has become prominent in the green technology and medical technology sectors as well. Wind turbines, solar cells, and batteries and engines for hybrid vehicles—key components of a future powered by alternative renewable energy—could not be manufactured without them. Advances in medical imaging have come about by the unique band gaps of elements such as gadolinium.

It seems that there are no limits to what the imagination can create with these elements, except that many of them are difficult to obtain. They occur unevenly across the globe, within geological deposits found in just a few countries. There is also specialized demand for one or another of them in any particular manufactured product. These elements hardly ever occur alone in pure veins. They are typically mixed with many other rare earths and other elements. Mining operations must often excavate large volumes of rock, often using open-pit mining that creates huge craters, merely to extract small quantities of these elements. Furthermore, the elements must be separated from the surrounding rock, and then separated from each other, using water, chemicals, and extraction processes that can be environmentally hazardous. The few desired elements become inputs to manufacturing; the remaining unusable or less useable ones are stockpiled as byproducts.

The rub here is that none of the desired elements can be fully substituted for one another, given the way current technology is fashioned. Our desire to have the latest technological gadgets therefore carries certain responsibilities and ethical obligations toward the environment. Our technological demand creates demand for elements from the entire periodic table that has global environmental consequences. As the Bristol Bay case example illustrates, this issue is often overlooked because we are far removed from the direct effects of the unhealthy environmental consequences of the extraction processes. Thinking about humans and nature entwined as socio-ecological systems means appreciating the growing, inextricable connectedness between global locations where technology is manufactured and used, and locations that physically provide the key elements. It is a poignant example of telecoupling.

Many of the key elements used in modern technology occur physically in geological deposits in wilderness areas like the Bristol Bay region. Decisions to mine in those highly valued, pristine areas are often met with strident protests. Actions to halt mining development are understandable based on a clear rationale to preserve these last vestiges of relatively untouched wilderness on Earth. Although this has long been regarded as an ethical position of high merit, the ethics of such a single-minded action has to be re-evaluated in a telecoupled world. Geological deposits are equally rare globally. Any contemplated action to halt mining in one cherished wilderness should be obligated to consider its global implications. In what other cherished wilderness in the world will the mining be done? What environmental damages and associated costs to human health and livelihoods will be passed on to those parts of the world?

Local-scale political tugs-of-war between wilderness preservation and mining address the issue at the wrong scale. Anyone who relies on modern electronic technology and encourages the development of green technology—environmentalists and technocrats alike—has a shared link to the global environmental impacts ensuing from mining in any one global location. An ethical position of environmental stewardship would obligate one to first question whether it is right to protect nature in one location and force resource extraction to be done in other parts of the world. Alternatively, if we do not wish to inflict damages elsewhere on the globe, would we be willing to forego the benefits of modern technology? Often the livelihoods of local residents whose employment depends on the existence of pristine wilderness (e.g., ecotourism), or of other residents who stand to gain the opportunity to work in the resource extraction industry, unwittingly hang in the balance. The link between technology and element extraction is a particularly clear example of the global linkages that create ethical and social conundrums that must be explicitly addressed and overcome if modern society is to achieve sustainability in a finite world with raw materials that are in limited supply.

Insights from the New Ecology offer lessons for reimagining how to deal with this frontline issue. Specifically, the field of industrial environmental management and engineering is increasingly borrowing ecological principles to inspire changes in how technology is designed, manufactured, and deployed. This new field, known as industrial ecology, advances a study and practice of engineering that helps to create a circular economy, by which I mean a manufacturing-based economy that is designed to minimize waste production, pollution, and environmental damages from resource extraction. A circular economy ensures that nutrients and materials translocate safely from industry to nature and reduces environmental toxicity by re-circulating metals within the industrial production system itself. Industrial ecology thus enhances society's ability to maintain the functionality of related ecological systems. With a goal to minimize the risk of damages without precluding opportunity, industrial ecology can, in many respects, be viewed as another form of preventive medicine for the environment.

* * *

Industrial ecology begins with the plain fact that societies build on their past technological base to advance new technology. It recognizes that society cannot sustain itself or improve without relying on technology. Industrial ecology proposes that societies can get inspiration from and can learn about sustainable

and efficient use of materials and energy by looking at how natural ecological systems function. Fundamentally, this means re-imagining that industries and society are integral players in circular economies that involve chains of producers, consumers and decomposers.

This view represents a significant departure from present-day conventions. Most industrial manufacturing effectively operates under a paradigm of a linear, or one-way, economy. Here, raw materials are extracted from nature, produced into value-added goods, sold to consumers in the market, and discarded once they exceed their useful life. In ecological vernacular, this is called an open system. Open systems can be sustainable only insofar as raw materials and energy are supplied in unlimited quantities. This condition will not be met for systems that depend on nonrenewable materials and energy (e.g., fossil fuels, and mineral elements). Appreciating this is especially important for finding ways to promote sustainable technologies.

Society nowadays largely promotes sustainability by expecting that technological innovations increase efficiency in the use of resources—materials and energy—that are in limited supply. This is accomplished by taking actions like improving fuel economy of automobiles that burn fossil fuels, or minimizing the wastage of raw materials during manufacturing. Such actions certainly increase the lifespan of those finite resources. Relying solely on efficiency, however, is insufficient because the resource will eventually be depleted anyway.

As supply of the desired resource dwindles, one might consider sequentially transitioning to the next limiting resource, and so on. But even if modern technologies could substitute one resource for another in the first place, this too cannot be sustained in the long term. Increasing the efficiency of resource use and substituting limiting resources are merely stopgap measures. They buy society time to transition to alternative, sustainable approaches. This then begs the question: What are the requirements for sustainability in systems that depend on finite resources? Industrial ecology addresses this question by looking at what sustains ecological systems.

An ecological system—a natural economy—that is dependent upon limiting resources will become sustainable when it becomes a closed system. Systems can become closed through feedbacks.

At a fundamental level, all spent and unused materials that are in limited supply must be returned back to the pool of available resources that support future production. This feedback is part of a process known in the ecological sciences as elemental cycling or materials cycling. Materials cycling requires decomposers. The decomposers that elemental cycling uses break down waste to its components for reuse in new production.

Society is abundantly familiar with this concept. It is called recycling. Societies actively participate by discarding all manner of products into recycling bins or other repositories. It has become a matter of course, and understood as a contribution toward environmental stewardship. Again, this is an ethical position of high merit. But perhaps less appreciated is that simply discarding products into recycling repositories does not guarantee that materials cycling will occur. It depends heavily on economic considerations, the costs and benefits of recycling relative to extracting new resources, which often do not favor recycling. The economics might be changed, however, if the environmental damages of resource extraction were factored into the calculus and, more importantly, if the recycling process itself were enhanced to make it more cost-effective.

As a process, recycling materials to keep pace with future demand depends directly on the decomposers' capacity to break down materials in the first place, and the speed at which materials can be broken down. However, many high-technology electronic products are routinely not built to allow them to be easily broken down into their elemental parts. As a consequence they become costly to break down, thereby straining the economic feasibility of recycling them. Often, they are merely left as discarded waste, collected in vast stockpiles, many of which can be toxic. This practice eats up land space and thereby can compromise the very ecosystems services that sustain human health and livelihoods: maintenance of clean water, food production, and gas regulation.

Environmental interest groups that wish to halt mining might do well by advocating for greater efficiency in decomposition. Ultimately, this means that recycling cannot be merely treated as an afterthought. Products must be developed and built with the deliberate, up-front intent to recycle them. Manufacturers of the products are therefore other appropriate targets of any advocacy aimed at minimizing environmental damages from mining, not just the miners.

Industrial ecology helps devise the key insights to make the process of materials use better able to support a sustainable, circular economy that can promote technological advances. It is all about planning ahead, by considering the full life of a potential product before it is even built. Such product design and engineering begins with the product's conception: deciding which materials will be used to build the working parts, the procedure for assembling the working parts, and most importantly ending with how the product will be taken apart and broken down once its useful life is over.

Such assessments provide the means to conduct scenario analyses, to weigh the pros and cons—efficiencies and financial and environmental costs—of using different materials and different manufacturing processes. This ensures that any new product is designed to be durable while in use. It further ensures that the product can be efficiently broken down into its elemental components when it is discarded. This helps to minimize the environmental damages from mining, and from industrial wastes and residues that end up in landfills.

Ecological principles also teach us that production and consumption process can be fine-tuned to maximize the flow rate and recycling of energy and materials through resource-limited, closed systems, i.e., to maximize sustainable economic activity. Recycling rates depend on how much of the limiting resource is available and how it is distributed among producers, consumers, and decomposers within society. When supply levels are low, producers cannot persist because there isn't a reliable enough supply to make production economically feasible. Increasing materials supply makes production economically feasible because increased, reliable production draws consumers. Consumers and producers can persist together as an economy. But consumers will reduce flows of limiting resources by holding onto the materials contained in the new products, rather than discarding them. That is, they increase the lifespan of the materials in use. Producing newer products encourages consumers to discard older products in favor of consuming the newer ones. But if consumers buy products faster than they can be recycled, the economy will again slow down because material is bound up in the products. The lifespan of the materials becomes too long. In other words, there is a point somewhere within the circular economy at which the flow of materials, and the associated financial activity related to it, could be optimized. Finding this optimum interval ensures that the flows of materials within the system can be quite large, and the flows into and out of the system that are due to leakage and inefficiencies are quite small.

Industrial ecology is taking an ecologically inspired approach to account for the distribution of elements within and among different humanly built systems. Such analyses calculate the stocks and the flows—the sources and fates—of elements at local to global scales. The analyses include elements like nitrogen and phosphorus used in agriculture, and metals like copper, aluminum, zinc, and nickel used in manufacturing. Such approaches are beginning to include some of the rare earths that will help stimulate modern technological advances. Current understanding is very rudimentary at best. So far, industrial ecologists are able to completely trace the cycles of only 30 of 103 elements of the periodic table on a global scale, and perhaps 45 within different subregions, countries, or territories of the globe.

Using nickel as a case example, analyses can already help to reveal where important gains and improvements can be made. For instance, the flows of nickel from mining to the use phase are large. This is because ore processing and smelting have become highly efficient to keep losses small. But the nickel being discarded is less than half of that coming into use. That is, much of the nickel is bound up in all manner of products. But at the other end, only 30 percent of the nickel that is turned into products comes from scrap recycling. Almost 50 percent of the discarded nickel ends up in landfills or held as general steel scrap. The remaining 20 percent is unaccounted for, but may be lost as leakage into the environment. Such analyses show that improvements at the input stage—mining extraction and processing—will lead to minimal gains. Larger gains could come from devising the means to efficiently recover nickel from manufactured products and redirect it back into manufacturing.

These analyses also reveal that there are treasure troves of elements bound up in existing and spent technology, and in human-built infrastructure. Much of it is not flowing. Moreover, given current trends, most of those bound-up stocks of elements merely accumulate. It turns out, then, that the mines of the future might well become urban areas, rather than wilderness, where metals are bound up in old buildings and infrastructure and landfills.

Achieving optimum flow of elements within cyclical economies depends on consumption but not on unbridled consumption. It also rests on the recycling rate of discarded materials for new production. If decomposition is inefficient, the economy will slow for two reasons: fewer new products can be manufactured; and consumers will hold on to the materials for longer periods of time. Or, put another way, decomposition, not consumption (consumerism), is what ultimately sustains the technological economy. It also reinforces the idea presented in chapter 5, that sustainable economies come from ensuring interconnections between important components of the system—producers, consumers, and decomposers—and the materials and energy flows between them remain intact. The manufactured goods then become the dividends of sustaining cycling and innovation within the whole system.

Ecological principles, especially the idea of food web networks, can also help re-imagine the structure and functioning of industrial facilities. Factories are essentially production systems involving inputs of unrefined materials, and outputs of finished products. In between, the materials are refined and transformed into components that are used to build the products. Such materials processing creates financially and environmentally costly residues, that are released as solid, liquid, and gaseous wastes to land, air, and water; or as heat. The solution is to configure industries into fully integrated networks, akin to food webs. This innovation takes a lesson from nature where one organism's waste is another's resource. In an industrial network, a group of industrial facilities collectively becomes an ecosystem—an industrial ecosystem—in which the members trade waste materials and energy amongst each other.

An example of such potential occurs in Denmark. Several altogether disparate facilities, including an electric power company, a pharmaceutical plant, a wallboard manufacturer and an oil refinery, exchange and use each other's steam, gas, cooling water, and gypsum residues. Other residues such as sulfur, ash, and sludge, are transferred out of the network. They are sold for use elsewhere, such as in sulfuric acid and cement manufacturing and farming operations.

This industrial ecosystem can also be considered self-organized in the sense that the whole ecosystem emerges from initiatives and negotiations of the participants themselves. It doesn't require politically or legally mandated actions. This emergent structure conceivably allows the network to flexibly change. It facilitates creativity and innovation by allowing different partners to forge linkages with existing or new partners, or dissolving them to create different ones, whenever any of the participant's economic fortunes change. The industrial ecosystem is thus not only self-organized, it also has the capacity to be adaptive.

Industrial ecosystems, like natural ecosystems, are not self-contained systems. They are highly likely to exchange materials and energy across their boundaries. But the nexus of exchanged materials between natural ecosystems usually does not cover great distances. The exceptions occur when materials from one ecosystem end up in another remote ecosystem, directed along by animal migrations or by prevailing wind or ocean currents. Industrial ecosystems are quite different because they are linked by many, many telecouplings that crisscross around the entire globe. This comes from global trade in all manner of raw materials, finished products, and everything between.

This global scale of exchange also raises the specter that trade in resources need not involve direct physical transfer of a resource from one location to another. It could involve trade in a virtual resource. The concept of virtual resource characterizes those resources that are used in the extraction or manufacture of traded products but not contained in the traded product. For example, countries that cannot manufacture their own products due to scarce resources, such as water used in mining and industrial manufacturing, may still benefit by trading with countries that have ample water supplies to support the manufacturing. In this case, the water is not traded directly because it isn't contained in the product. It is a virtual resource because it is used to create the product. The concept of virtual resources helps provide a full and explicit accounting of all the resources needed to make different products. Such accounting can enhance environmental stewardship in a telecoupled world by revealing how strain on the environment and human livelihoods arise in one location driven by demand for manufactured goods in another.

Industrial ecology is also now toying with using principles of evolutionary biology and resilience. The capacity to evolve rests on the freedom to innovate with how materials are processed and exchanged. Such capacity can help to build resilience in the face of changeable human demands and desires, and economic shocks. The capacity to be resilient rests on the ability to resist the effects of a shock to a system, or to recover very quickly to resume normal function once the shock abates. Resilience also embodies the idea that systems can flip into alternative states. Accordingly, a resilient system is also one that averts shifts to alternative states by responding nimbly and adaptively to disturbances.

But there are upsides and downsides to making industrial facilities or industrial ecosystem resilient. It depends upon whether one wants to keep a system of interest in a specific state or switch to a more desirable alternative.

Resilience can cause much resistance to change when such change is crucially needed. For example, modern society is locked into production economies that are supported largely by energy derived from fossil fuels. Moreover, emerging technological societies choose to operate within this fossil-fuel driven state by investing in more effective ways to extract the ever-dwindling supplies of fossil fuels. The alternative is to begin anew by transitioning toward clean and renewable energy technology. But the lack of social or economic will or ability to innovate and evolve to renewable energy state is what makes the existing fossil-fuel state highly resilient.

Alternatively, systems can be locked into undesirable states despite willingness or need to evolve, simply because players within the system cannot change fast enough. The North American automotive industry has been a clear example. The industry was almost singularly geared to build large vehicles with high fuel consumption. In the face of abrupt and rapidly rising fuel prices the industry had a poor capacity to adapt to sudden changes in consumer demands for vehicles that have greater fuel economy, or ones that use alternative energy (hybrid vehicles). Consequently, the industry was vulnerable to collapse, losing out to foreign competitors that provided such choices in vehicles. That is, North American manufacturers lacked evolutionary capacity—the diversity of vehicle models that could be selected by consumers—to overcome the undesirable state. By becoming too specialized—too brittle—the industry painted itself into a proverbial corner by reinforcing the now undesirable state and losing the adaptive capacity to escape it. Building adaptive capacity might involve having a stable of new product innovations ready to be rapidly implemented in anticipation of the emergence of a different economic climate. Or, it could include the capacity of design teams to innovate products and the manufacturing processes and to rapidly implement those new innovations.

The challenge in maintaining or transitioning to desirable sustainable states is that uncertainty about future conditions makes it difficult to decide what strategies or processes to maintain. However, from an evolutionary ecological perspective a key strategy is to maintain at all times the capacity to innovate and create quickly. This capacity can be built into systems at several levels of a hierarchy. Sudden, small shocks might be accommodated by encouraging capacity for rapid adjustments within the day-to-day operations of a manufacturer. This comes about by maintaining the ability to fine-tune the efficiency of production streams in the face of small jumps in the price of materials or energy. Intermediate shocks, such as shortages in certain materials or energy sources may require strategic initiatives that alter how products are manufactured. Large shocks such as demand for radically different kinds of products may cause a particular state to collapse entirely and in turn require wholesale change in the way an industry operates. This "destruction" provides new creative opportunity. It can lead to evolutionary change through selection for capacity to provide entirely new ways of manufacturing products and ultimately in the new products that are developed. We see this all the time through rapid changes in technology and the companies that produce them. The rapid change in home entertainment from VCR, to DVD, to pod technology, to on-line streaming is but one example. It also may mean wholesale changes in corporate strategies. For example, it may require that strongly individualistic companies or competitors organize themselves into industrial ecosystems by forging collaborative and cooperative interactions that are built on fragile trusts among the interacting partners.

The point here is that striving to create resilient systems is customarily looked upon as a socially and environmentally responsible action to achieve the sustainability goals of human well-being and environmental health. The concept of resilience and alternative states, when applied to industrial systems, teaches that there is a counterintuitive downside to all of this. The New ecology teaches us that any tendency for humans to try to hold on to an existing system, because they fear the unknown and the possibility that their health and economic well-being will be imperiled, can lock socio-ecological systems into undesirable, unsustainable states.

* * *

The notion that large urban areas may become the mines of the future is a direct consequence of a sea change in global political geography. Urbanized areas are those that are densely populated by humans, and dominated by human-constructed features like buildings for habitation and work, and infrastructure for transportation. Cities represent an extreme form of an urban setting in which humans tend to be very densely packed into small areas.

In the early 1800s when the global human population was around a billion people, there was only one city—Beijing—whose population exceeded a million inhabitants. A century later, there were 16 cities worldwide that had that number of inhabitants or more. Currently that number is 200 and is slated to soar to 600 by 2025. Moreover, for the first time ever, better than 50 percent of the global population is concentrated into cites rather than spread across landscapes. Demographers and geographers expect that, by 2100, the fraction of the global population dwelling in cities will reach 70 to 90 percent as more individuals move out of the surrounding countryside. Remarkably, however, cities currently occupy a mere 1 to 2 percent of the available global land space. They may not occupy very much more than that in the future.

So, in some sense, one might be inclined to argue that this redistribution and concentration of global human populations into less land space is a boon for wilderness, right? The future may in fact hold great promise for protecting nature, right? In fact, we don't need to worry about overcoming the human/nature divide, right? Not so fast.

Part of the issue is that human populations are still growing in these areas. Influxes of migrants into the cities, as well as births within them, swell urban population sizes. Such population growth amplifies the demand for resources. But there are fewer and fewer prospects for resource production and extraction (e.g., food, energy) within the city limits because land space is continually being re-allocated to urban infrastructure. At a fundamental level, then, these urban areas—urban systems—are being designed and developed as open systems. This means that they will become unsustainable systems. The New Ecology can offer guidance to ensure that the growing trend toward urbanization doesn't create unstable human-built environments.

Cities currently draw most of their resources from outside of their boundaries. They use 60 percent of available freshwater from across landscapes, and they utilize 70 percent of wood available for housing and industrial activity. Seafood demand in the 740 largest cities globally appropriates 25 percent of the all the marine production coming from marine shelf, coastal, and upwelling areas.

All this resource consumption fuels activity within city boundaries that helps to raise humankind from poverty. Cities collectively are responsible for 90 percent of global gross domestic product (GDP). As a consequence, urbanized populations tend to be more affluent than rural populations. This leads to changes in diets, in favor of greater meat consumption, and increased demand for the latest manufactured goods and technology. Thus, an increasingly urbanized human society continues to place demands on resources such as mineral elements, or wood fiber, or food. So, pressure on wilderness and less urbanized rural areas will not be abated. In fact, the pressure may increase. Estimates suggest that, given current trends, the space needed to support resource demands for urban areas may need to be quadrupled. And, given current flows of trade in resources and manufactured products, that pressure will be felt all across the entire globe.

Cities also have large outputs. Fully 70 percent of global carbon emissions to the atmosphere come from the generation and use of energy for manufacturing, heating and lighting, and transportation. Cites globally also produce 1.2 billion metric tons of solid waste each year. This mass is equivalent to about 3,500 of the largest ocean-going container ships (each the size of four football fields) when there are fully loaded. The amount of waste production is expected to triple by 2100, if current trends continue.

But current trends need not continue. If we can reinvent engineering principles to develop closed-loop, sustainable industrial processes, surely we can scale-up those same principles to close the loop and create sustainable cities or urban systems. The emerging field of urban ecology is applying ecological principles to help reorient urban planning do just that, and more.

Urban ecology studies patterns of building and infrastructure development, resource and materials flows, and environmental impacts, across space and over time. In essence, it involves the kinds of systems-based analyses of materials and energy flow used in industrial ecology but writ over larger spaces. And, it perhaps goes without saying, it concerns itself with how human socio-economic and political geography influence the stocks and flows of materials and energy. But it does more than that. It tries to align those socio-economic and political considerations with biodiversity conservation, in order to sustain or restore ecosystem functions. The aim is to design cities so they become more self-sustaining. It devises ways to build in elements of nature to enhance environmental services from ecosystem functions operating within city limits. This certainly supports burgeoning urban populations there. But it also helps to reduce demands and pressures placed on locations outside the city limits. The urban watershed and greenspace (see chapter 3) is one example of the kind of urban planning that can help cities promote biodiversity in ways that maintain environmental quality in cost-effective ways, while decreasing hindrances to those living beings outside of the immediate urban realm. As such, it is founded on an environmental stewardship ethic. It helps humans meet their ethical obligations to each other and nonhuman organisms that are mediated through their mutual relationships with the environment.

Like Biosphere 2, urban areas are effectively engineered places that contain mosaics of ecosystem types. They include human-engineered ecosystems like tree-lined streets, lawns and parks, urban forests, and cultivated land. But, unlike Biosphere 2, these ecosystem types are mixed in with, and thus complemented by, natural ecosystems like wetlands, lakes and sea, and streams. Together, these mixtures of ecosystems types have the potential to offer many environmental services. These include air filtration, rainwater drainage, and sewage treatment.

Take urban trees as a case. Air filtration services by urban trees may help remove pollutants like ozone, nitrogen, and sulfur dioxides, and small particulates. Trees offer the added benefit of cooling urban areas through shading. This service in turn prolongs the life of infrastructure like paved roadways. It also helps save energy that would normally be used to cool buildings. This in turn reduces emissions of greenhouse gasses and particulate air pollutants that accompany energy generation. Rooted trees help stormwater percolate into soils rather than run-off across impervious surfaces and flooding urban drainage systems and watercourses, carrying with it pollutants from city streets. This helps to protect water quality in urban watersheds. Estimates indicate that the value of these services to a given city could amount to hundreds of thousands to millions of dollars, depending on the size of the urban area. The replacement value of the trees can reach hundreds of millions of dollars. These sums reflect net benefits: they account for management costs like tree planting, pruning, and removal, leaf pick-up and disposal, and utility-line clearing.

Urban trees offer personal health benefits as well. People living in neighborhoods with high densities of roadway trees are characterized as having higher perceptions of personal physical and mental health, of feeling younger, and of having lower incidence of cardiac and metabolic ailments than people living in the same city but in neighborhoods with fewer trees. It also encourages people to eat healthier diets, especially less meat and more servings of vegetables, fruits and grains. These health indicators persist even after accounting for differences in socio-economic factors and age. It is estimated that these lifestyle effects are equivalent to having $10,000 more in personal annual income and moving into neighborhoods whose residents have $10,000 more average annual income.

Still, urban environments are domesticated nature in the extreme. So there are numerous downsides. Domestication decreases habitat for nonhuman species, and any habitat that remains is often highly fragmented. Urbanization can change the composition of ecological communities by selecting for those species that are better able to thrive in humanly dominated settings. Urbanization also can alter food webs and the degree of top-down and bottom-up control over the flow of nutrients within and among the mosaic of ecosystems. It can change hydrological cycling and stream flows through increased water use, water contamination, impervious surfaces, altered runoff patterns, and modified evapotranspiration rates. Urban soils are often physically disturbed, chemically contaminated, or compacted by management practices.

The scientific understanding of these effects still remains rudimentary. A growing, concerted effort in urban ecology has begun accurately measuring the various stocks of materials and energy within the urban system and tracing their flows into, within, and out of urban systems. Such analyses will improve understanding of the benefits as well as costs of implementing ecological principles. Using ecological principles, informed by newly gathered scientific evidence, can abet urban planners to think more creatively about building sustainable urban environments.

Some of that creativity may come from combining natural features, like plants having different structures, in construction processes to enhance the service value that urban nature can offer. Green roofs, roofs of buildings that are covered by vegetation in a growing medium, are one such example. Green roofs help absorb rainwater, provide insulation, and lower urban temperatures to mitigate urban heat island effects. Bioswales are another example. These gently sloping landscaping elements create a drainage course—a modified ditch or local depression, if you will—that is filled with natural vegetation or compost. They are usually built along streets or in parking lots. The swale collects and holds water from surface runoff to filter out silt and pollutants before the water eventually enters a city stormwater sewer system, and thereby the swale protects urban water quality. These innovations herald a new era for designing and building urban systems. There is much opportunity to experiment—to take a scientific approach to urban design—by testing out the efficacy and performance of different ecologically informed construction features, and then refining them at small scales, before they are implemented wholesale at the city level.

* * *

Even with the best of intentions, urban planners will be unable to achieve fully sustainable cities by closing the loop within the confines of the cities themselves. The land space available for say even just food production, will not come anywhere close to having an economy of scale needed to support all of the inhabitants of any one city. Closing the loop thus requires looking beyond the city boundaries and understanding the stocks and flows of critical resources and wastes that come into and go out from the cities. Achieving sustainability requires understanding how any one city is telecoupled to other locations globally.

Take for instance the production of meat (chicken and pork) for consumers in Japanese cities. This demand is telecoupled to production systems in China, Brazil, and the United States. About two million hecatres of land—the same area as the city of Tokyo—are needed to produce feed crops for these animals. In addition, crop production requires about 35 cubic kilometers of water each year. This is equivalent to about three Lake Meads, the largest water reservoir in the United States, formed by the construction of the Hoover Dam.

The intimate connections between urban regions that receive materials, energy, and products from geographic regions that produce and send them, means that the geographic extent of a city's impact goes far beyond its immediate boundary. Efforts to reimagine this geographic extent, in a functional sense, can help to close the loop and make cities meet ecological requirements for sustainability. Urban ecologists are developing the means to do this through the concept of systems integration.

Systems integration aims to overcome the fact that the many causes and consequences of human impacts on the environment too often stem from a failure to look beyond the boundaries of any one system. Achieving urban sustainability requires an accounting of how the effects in one urban system are connected to and feed back from other systems in a telecoupled world. It is an issue of complexity to be sure. The accounting must consider the flows of materials and products within and between the coupled systems as well as the social, political, and economic factors that shape those flows. It means using that accounting to find critical levers that can produce significant gains in sustainability. It involves follow-up accounting to see if the practices are achieving their goals.

Consider a case example from China to see how system integration might come about. The city of Beijing does not have sufficient water supplies within its boundaries to support the needs of its twenty million inhabitants. It has to look 100 kilometers away to the Miyun River watershed to source its freshwater. The water reservoir built there is the largest artificial lake in all of Asia. It is about a tenth the size of Lake Mead. This source area is also important to the livelihoods of the 878,000 residents in the watershed, 97 percent of whom are engaged in agriculture. Building a reservoir to hold freshwater there certainly benefits Beijing residents, but it precludes opportunities for those residing in the watershed. One solution, then, is to pay residents in the source area for the ecosystem service—provisioning of freshwater—that they provide to Beijing. However, this solution by itself does not ensure that the residents can carry on with their traditional agrarian life styles, which is furthermore important to sustaining the social fabric of the local communities. On the other hand, local residents cannot continue their traditional agrarian practices because they stand to compete for water yields and pollute water quality after nitrogen and phosphorus leach from the fertilizer applied in order to grow crops.

The solution that the city of Beijing adopted is to develop a regional collaboration with these agricultural communities to deliver clean water in abundance. Beijing incentivizes this collaboration by tying payments for freshwater provisioning to agricultural practices that protect the watershed and hence protect water quality and quantity. Specifically, rice paddy agriculture demands a lot of water and can be highly polluting from high nitrogen and phosphorus loading in the water. The solution is to use fewer water-demanding crops like corn. Moreover, corn cultivation is aligned with appropriate landforms and soils in the area. Environmental analyses show that it reduces the amounts of nitrogen and phosphorus that end up being leached into the reservoir.

Follow-up accounting shows that the agricultural practices have altered social dynamics and household economies as well. Households have greater income, but less of it comes from the less lucrative corn agriculture and more comes from nonagricultural migrant work. Households also change their spending habits with their greater ability to buy technology and to afford education. They reduce the use of fuel woods that come from the watershed, but they increase the use of coal and petroleum. Understandably, these changes can have feedbacks to other sectors of the environment. But aligning economic activities explicitly with environmental goals through increased coordination within the Beijing and Miyun socio-ecological systems represents an important step in the direction of creating flexibility in policy development and action. Follow-up analyses help identify gaps in knowledge, technology and governance that must be addressed to reach their full potential for those integrated socio-ecological systems to become sustainable. The key here is that flexibility can help the integrated systems adapt in ways that fine-tune their environmental performance.

The Beijing-Miyun coupling is an example of systems integration in space. There are, however, integrated systems that have socio-ecological implications that are realized over time. A clear example is the conundrum over using agricultural crops to feed humans versus using them as a source of energy in the form of biofuels. Many countries across the globe are highly dependent on industrial-scale crop production from a few countries. But for reasons of global security, or perhaps to reduce environmental impacts of fossil-fuel burning, some producer nations may redirect their crop production to the manufacture of biofuels. The potential net environmental costs of this practice notwithstanding, policies that encourage rapid re-allocation of food crops for biofuels production can send shocks through the telecoupled geographic locations. This imposes future stresses on communities that are dependent on trade of food crops because they are not resilient. They have limited means to feed themselves from local production and have limited capacity to adapt by increasing local agriculture. This includes many developing nations, but includes developed nations like Japan or the Netherlands as well, given their current land allocations for nonagricultural uses. Systems integration helps to understand and anticipate that responding to one security issue could create another security issue at a future time.

Industrial and urban ecology are emerging fields that represent the culmination of everything discussed in the book so far. They are based on systems thinking and include ecosystem service valuation, considerations of planetary boundaries on capacities of ecosystems to function and provide services, telecoupling of real and virtual resources, the hierarchy of human social systems comprised of social, political, and economic organizations and institutions, and environmental stewardship. They offer society ideas about entwining humans and nature to achieve sustainability in ways that are respectful and ethical to both.

CHAPTER 8

THE ECOLOGIST AND THE NEW ECOLOGY

For ecologists advancing the New Ecology, the ultimate penalty of having an ecological education is the personal struggle we face between being neutral and being invested in the cause of sustaining nature and humanity. The struggle requires finding a happy personal balance between being engaged in a vocation that involves studying the intricacies of nature to advance scientific knowledge, driven by fascination to understand its mysteries; and advocating out of passion to protect nature in the very interest of sustaining both nature and society. Working in a science that is increasingly being looked upon for help to achieve sustainability, we shouldn't want ecologists, as professionals, to do things any other way. Ecologists got to where they are today by rethinking their research approaches and their theories, and thereby refining their scientific understanding of nature and of humanity's role in it. It has changed ecologists' own view of the role they should assume in informing society.

In its beginnings, ecology was effectively a science in support of sustaining nature for itself. During this period, scientific analyses were focused on developing a basic understanding of species and ecosystems, informed and motivated by natural history observations of species in their localized habitats. The dividend of this enterprise was that it had amassed the scientific knowledge and ability to identify and characterize species and ecosystems at risk. It began an era of environmental conservation and of science-based natural resource management. Ecological scientific insights were used to inform advocates for conservation of species or of land and advocates against the destructive overexploitation of nature by humans.

The rise in human population growth and the increasing human domination of nature throughout the twentieth century largely caused ecologists to redouble their efforts, to grow ecology to become a science in support of sustaining nature despite people. Ecologists devised experiments that revealed the ecological consequences of changing presences and abundances of species, and the configurations of their habitats. It refined ways to characterize and pinpoint species and ecosystems at risk. It was a time when the science motivated many efforts globally to designate and protect large swaths of wilderness as national parks and protected areas. Ecologists helped to inform policymakers how large these parks and protected areas ought to be, how many there ought to be across a region, how connected they ought to be, and how representative of nature they ought to be. But this effectively helped to promulgate a human/nature divide.

Humanity, however, continued to encroach on wilderness. In many places across the globe, conservation based on a view of that divide displaced people from their traditional lands. It thereby disrupted their communities, causing them to lose their sense of place, their livelihoods, and perhaps even the value they held toward nonhuman life. As a consequence, it shifted human usage of landscapes, and reorganized the locations and scales of human impacts. In many cases, this led perversely to the kinds of "islands" of protection and threats to biodiversity and sustainability that I earlier described (see chapter 4). But the tables were also turning with the rising scientific appreciation that humankind valued nature's material benefits and services to support their livelihoods, sometimes, perhaps, as much as or more than they valued the mere existence of nature. This gave rise to the New Ecology in which scientific analyses focused on how nature provided those services. Ecology thus grew to become a science in support of sustaining nature for people. It led to the kinds of research that I have been intent on describing. This fresh scientific role is certainly helping to overcome the human/nature divide by promoting the view that nature—biodiversity and ecosystem functions—must be protected to provide the suite of environmental services, inside as well as outside of protected areas, in support of humanity. But ecologists also have become uneasy about the implications of that science, because it advances a largely utilitarian, human-centric view of humanity's engagement with nature.

To be an effective approach for human engagement, ecologists needed to articulate the message that sustaining nature for people was more than a one-way street. As members of a working social/economic/political/ecosystem, humans have a shared responsibility to sustain nature's "well-being" in addition to the well-being of one another through their mutual engagement with nature. What was lacking, however, was an ethical basis for framing such engagement. This led to the idea of environmental stewardship, a science-based ethic that aims to guide human action by viewing humans and nature entwined as a socio-ecological system. Ecology—the New Ecology for the twenty-first century—has thus become a science in support of sustaining humans and nature in the Anthropocene.

The job, however, is certainly far from complete. Ecologists are now heavily engaged to improve knowledge creation by advancing a big-picture science that is commensurate with the geographic scales of humanity's reach. Humans will inevitably be the predominant forces shaping the world during the next century. With this comes the looming worry that Earth's space will be rapidly transformed and repurposed, leading to widespread loss of biodiversity and ecosystem functioning. This certainly has become an appealing conventional wisdom. But the scientific evidence on biodiversity trends suggests that the emerging picture may not be as straightforward as that.

Ecologists are now responding to the call to assess and map global trends in biodiversity, to gauge what the magnitudes of loss might be. This is certainly viewed by ecologists as an important and fundamental enterprise that would advance the science of ecology. Yet, it is now also viewed as an important contribution to achieve more than that. Ecologists are also realizing that understanding and predicting global trends in biodiversity has important implications for sustainability, ethics, and environmental policy in the interest of both humans and nature.

A hard look at biodiversity trends across the entire geography of the planet is revealing new insights that question the conventional wisdom of widespread biodiversity loss. For example, many natural places like oceanic islands that have often been deemed fragile to human impacts have in fact not lost biodiversity. Inventories based on counts of species on islands show that in some places biodiversity has even increased, despite documented extinctions that have resulted from human invasion of these places. Newer inventories are beginning to show that local and regional biodiversity, when examined across continents, may be remaining steady or increasing. At the same time, such analyses are beginning to highlight that examining trends, simply using counts of species as an all-encompassing measure of biodiversity, may not provide a complete picture of the state of affairs.

For example, the total biodiversity in many geographic regions may not have changed much, even though the species that once occupied those locations are no longer there, replaced by other colonizing or invading species. This phenomenon of species turnover—where species that have gone extinct within locations are replaced by new species—is brought about in part because humans have altered and rearranged landscapes to make nature more rambunctious. But the data are still very sketchy on how all of this is playing out.

Abetted by modern remote sensing and geographic information systems technology, ecologists are beginning to characterize and map out this new rambunctiousness. Ecologists can now resolve the spatial patterning of global land use changes and the temporal re-arrangements of species concentrations across vast geographic spaces. Ecologists are developing newer ways to measure the patterns in distributions and abundances of species and thereby to coordinate and deploy concerted efforts and methodologies to gather new data that will provide coherent inventories across the globe. In effect, they are promoting a new natural history for the digital age. In many ways, it is not unlike classic, place-based natural history, in that it provides detailed descriptions of what we see in nature. But it has become more sophisticated quantitatively and broader in scope. It is amassing large, global-scale datasets, and developing and applying new statistical procedures that characterize and map changes in associations between species and their habitats across broad geographic scales. It is thereby advancing a new dynamic picture of species re-arrangements. It raises the possibility—the hypothesis that needs further scientific testing—that many species may be more resilient, more capable of coping and adapting to human domestication of nature, than conventional wisdom has long held.

Characterizing which species are where is of course only half the story. The other half is telling what the particular arrangements within geographic locations means for ecosystem functioning and services. Ecologists want to know whether, if one species replaces another, like is replacing like functionally. That is, does species turnover mean that ecosystem functioning also changes; or do the new species assume the role once held by the species they have replaced?

The New Ecology has made important advances in explaining the functional roles of biodiversity within ecosystems based on species identities and their functional traits. That understanding was built using experimental approaches that required communities and food webs to be assembled randomly. This was deliberate according to the rules of traditional experimental design, to avoid biasing the experimental findings. But human domestication of nature does not affect biodiversity randomly. For instance, as we have come to learn, predator species across the board, from vertebrates like mammals and fish to arthropods like insects and spiders, are declining in abundance or geographic presence disproportionately faster than other species. Ecologists are now confronting the emerging new question of what such nonrandom changes in species composition of ecosystems means for ecosystem functioning. Along with this comes an important challenge to devise new experimental approaches that can speak to issues playing out at the large spatial extents and long time periods that are commensurate with the scales of human engagement with nature. This challenge is conflated especially by the fact that humans do not engage with nature in nicely controlled and random ways.

With this also comes the growing realization that the resilience of ecological systems may depend in large degree not just on the species comprising ecosystems, but also on the diversity of evolutionary heritages of the species—called phylogenetic diversity. Ecologists are working to characterize patterns of phylogenetic diversity in space and time and relate it to functioning by probing how an evolutionary heritage that shapes species traits determines how species fit together functionally in ecosystems. It is raising all kinds of questions about what it would mean for sustainability if society were to sever these evolutionary heritages through changes humanly wrought across landscapes.

In the most interesting turn of events, ecologists are now gravitating toward studying nature in ecosystems created and heavily populated by humans. Humanity calls them "cities" and "urban areas"; but ecologists see them as another kind of ecosystem, albeit one with no direct wilderness analog. Urban areas are unique landscape mosaics that arise from interspersing built infrastructure among the natural features of the land, its vegetation, and its geomorphology. This new landscape changes the ecosystem's microclimate, flows and concentrations of water and nutrients, and emissions and concentration of pollutants. Planted landscaping can create habitats for nonhuman species. But the exact kinds of plantings often can act as filters, attracting only a subset of species, those that are capable of living in those constructed habitats. Many, many more species may be excluded. By design, the mosaic nature of urban landscape development encourages habitat fragmentation. Urban transportation networks and the associated bustling flow of traffic can exacerbate this fragmentation and it can risk extinction of species by disconnecting those habitats. With the intent to improve urban design and planning, ecologists are beginning to research what all this disconnectedness might mean for the functionality of urban ecosystems.

Urban ecosystems also offer new testing grounds to see how well modern ecological scientific principles will hold up to explain the workings of twenty-first-century nature—a domesticated nature that includes wilderness, densely populated urban environments and everything in between. The extent and pace of urbanization raises the specter that species will face new and stronger evolutionary pressures. An emerging research frontier is to understand the role that humans play in eco-evolutionary processes, how they instigate evolutionary change, and what any evolutionary change means for ecological functionality and sustainability.

The specter that humans can instigate rapid evolutionary change is well appreciated in an environmental stewardship ethic. It is embedded in the concept of complex adaptive socio-ecological systems. But what it means operationally for the interplay between changes in ecological systems and social institutional change remains beyond current comprehension. Moreover, what does it mean ethically to have an important hand in driving a creative process that has given rise to the dizzying variety of life we see today, and one that will shape what we see in the future? What is humanity's obligation to ensure that evolutionary capacity—central to ensuring resilience—is sustained? It is humbling, even to think about all of this.

The realization that humankind can have a profound role in shaping what the Earth looks like in the coming century creates a sense of urgency to advance new scientific principles that can answer the growing list of new questions about how socio-ecological systems function. Ecologists cannot, however, go it alone any longer. It requires advancing creative new ways of blending the study of ecology and the study of humanity. It means overcoming a new human/nature divide, one that comes from scholars steeped in the norms and conceptual views held within their separate scholarly traditions of ecology, economics, social science, political science, philosophy, theology, geography, among others. Ecologists are assuming leadership roles in encouraging more integration of scholarly fields, in order to develop an expanded conceptual view of what it means to be environmental stewards (or if that is even an apt way to view what humanity's role should be) going forward.

It is an exciting time to be an ecologist. Humanity's influence is forcing us to stretch our imaginations into many realms that we have never considered before, both in our science and the application of our science. By helping humankind to better distinguish its wants from its needs, and to understand the ecological as well as the societal implications of both, ecologists are helping societies transition toward more sustainable livelihoods. To this end, ecologists in the New Ecology are beginning to live up to the ideal called for long ago by Aldo Leopold. The New Ecology that I have here described is one that sees itself as a science in support of sustaining the amazing variety of all life—humans and nonhumans alike—as fully integrated nature. To this end the New Ecology is devising the scientific means to practice curative as well as preventive medicine for this new human/nature intertwinement.

BIBLIOGRAPHY

CHAPTER 1

Aber, J. D., and Jordan, W. R. 1985. Restoration ecology: An environmental middle ground. *BioScience* 35 (7): 399.

Costanza, R., d'Arge, R., de Groot, R., Farber, S., Grasso, M., Hannon, B., Limburg, K., Naeem, S., O'Neill, R. V., Paruelo, J., Raskin, R. J., Sutton, P., and van den Belt, M. 1997. The value of the world's ecosystem services and natural capital. *Nature* 387: 253–260.

Cronon, W. 1995. The trouble with wilderness; or, getting back to the wrong Nature. In *Uncommon Ground: Rethinking the Human Place in Nature* (W. Cronon, editor). Norton, New York.

Daily, G. C. 1997. *Nature's Services: Societal Dependence of Natural Ecosystems*. Island Press, Washington, DC.

Frumkin, H. 2001. Beyond toxicity: human health and the natural environment. *American Journal of Preventative Medicine* 20: 234–240.

Kareiva P., Watts, S., McDonald, R., and Boucher, T. 2007. Domesticated nature: Shaping landscapes and ecosystems for human welfare. *Science* 316: 1866–1869.

Koellner, T., and Schmitz, O. J. 2006. Biodiversity, ecosystem function and investment risk. *BioScience* 26: 977–985.

Leopold, A. 1953. *Round River.* Oxford University Press, Oxford, UK.

Marris, E. 2011. *Rambunctious Garden: Saving Nature in a Post-Wild World*. Bloomsbury, New York.

Myers, N. 1996. Environmental services of biodiversity. *Proceedings of the National Academy of Science USA* 93: 2764–2769.

Norton, B. 1996. Change, constancy and creativity: The new ecology and some old problems. *Duke Environmental Law and Policy Forum* 7: 49–70.

Power, M. E., and Chapin, F. S. III. 2009. Planetary stewardship. *Frontiers in Ecology and the Environment* 7: 399.

Ryff, C. D., and Singer, B. 1998. The contours of positive human health. *Psychological Inquiry* 8: 1–28.

Schindler, D. E., Armstrong, J. B., and Reed, T. E. 2015. The portfolio concept in ecology and evolution. *Frontiers in Ecology and the Environment* 13: 257–263.

Schmitz, O. J. 2007. *Ecology and Ecosystem Conservation*. Island Press, Washington, DC.

Seeman, J. 1989. Toward a model of positive health. *American Psychologist* 8: 1099–1109.

CHAPTER 2

Bonakdarpour, M., Flanagan, B., Larson, J., Mothersole, J., O'Neil, B., and Redman, E. 2013. The economic and employment contributions of a conceptual Pebble Mine to the Alaska and United States economies. HIS, Inglewood, CA. http://corporate.pebblepartnership.com/files/documents/study.pdf.

Botkin, D. B. 1990. *Discordant Harmonies: A New Ecology for the Twenty-first Century*. Oxford University Press, Oxford, UK.

Carlson, M., Wells, J., and Roberts, D. 2009. The carbon the world forgot: Conserving the capacity of Canada's boreal forest region to mitigate and adapt to climate change. Boreal Songbird Initiative and Canadian Boreal Initiative, Seattle WA, and Ottawa. 33 pp.

Centers for Disease Control and Prevention, Division of Vector-Borne Diseases. 2013. West Nile Virus in the United States: Guidelines for Surveillance, Prevention, and Control.

Duffield, J. W., Neher, C. J., Patterson, D. A., and Goldsmith, O. S. 2007. Economics of wild salmon ecosystems: Bristol Bay, Alaska. USDA Forest Service Proceedings RMRS-P-49.

Falkowski, P. G., Fenchel, T., and Delong, E. F. 2008. The microbial engines that drive Earth's biogeochemical cycles. *Science* 320: 1034–1039.

Fang, J. 2010. A world without mosquitos. *Nature* 466: 432–434.

Jorgensen, S. E., Fath, B. D., Bastianoni, S., Marques, J. C., Müller, F., Nielsen, S. N., Patten, B. C., Tiezzi, E., and Ulanowicz, R. E. 2007. *A New Ecology: Systems Perspective*. Elsevier, London.

Keohane, N. O., and Olmstead, S. M. 2007. *Markets and the Environment*. Island Press, Washington, DC.

Malaj, E., von der Ohe, P. C., Grote, M., Kühne, R., Mondy, C. P., Usseglio-Polatera, P., Brack, W., and Schafer, R. B. 2014. Organic chemicals jeopardize the health of freshwater ecosystems on the continental scale. *Proceedings of the National Academy of Science USA* 111: 9549–9554.

Norton, B. 1996. Change, constancy and creativity: The new ecology and some old problems. *Duke Environmental Law and Policy Forum* 7: 49–70.

Peterson, R.K.D., Macedo, P. A., and Davis, R. S. 2006. A human-health risk assessment for West Nile virus and insecticides used in mosquito management. *Environmental Health Perspectives* 114: 366–372.

Relyea, R. A., and Diecks, N. 2008. An unforeseen chain of events: Lethal effects of pesticides on frogs at sublethal concentrations. *Ecological Applications* 18: 1728–1742.

Rose, R. I. 2001. Pesticides and public health: Integrated methods of mosquito management. *Emerging Infectious Diseases* 7: 17–23.

Schmitz, O. J., Raymond, P. A., Estes, J. A., Kurz, W. A., Holtgrieve, G. W., Ritchie, M. E., Schindler, D. E., Spivak, A. C., Wilson, R. W., Bradford, M. A., Christensen,

V., Deegan, L., Smetacek, V., Vanni, M. J., and Wilmers, C. C. 2014. Animating the carbon cycle. *Ecosystems* 7: 344–359.

Swimme, B. T., and Tucker, M. E. 2011. *Journey of the Universe*. Yale University Press, New Haven, CT.

CHAPTER 3

Cardinale, B. J., Duffy, J. E., Gonzalez, A., Hooper D. U., Perrings, C., Venail, P., Narwani, A., Mace, G. M., Tilman, D., Wardle, D. A., Kinzig, A. P., Daily, G. C., Loreau, M., Grace, J. B., Larigauderie, A., Srivastava, D., and Naeem, S. 2012. Biodiversity loss and its impact on humanity. *Nature* 486: 59–67.

Daily, G. C. 1997. *Nature's Services: Societal Dependence of Natural Ecosystems*. Island Press, Washington, DC.

Dirzo, R., Young, H. S., Galetti, M., Ceballos, G., Isaac, N.J.B., and Collen, B. 2014. Defaunation in the Anthropocene. *Science* 345: 401–406.

Hansen, A., Peterson, J., Ellis, J., Sedneck, G., and Wilson, B. 2008. Terrestrial and aquatic linkages: Understanding the flow of energy and nutrients across ecosystem boundaries. http://classes.warnercnr.colostate.edu/fw400/files/2011/09/FW400-Terr-Aquat-Linkages-magazine-2008.pdf.

Harper, J. L. 2010. *Population Biology of Plants*. Blackburn Press, Caldwell, NJ.

Horowitz, P., and Finlayson, C. M. 2011. Wetlands as settings for human health: Incorporating ecosystem services and health assessment into water resource management. *BioScience* 61: 678–688.

Koellner, T., and Schmitz, O. J. 2006. Biodiversity, ecosystem function and investment risk. *BioScience* 26: 977–985.

Loreau, M., Naeem, S., and Inchausti, P. (editors). 2002. *Biodiversity and Ecosystem Functioning: Synthesis and Perspectives*. Oxford University Press, New York.

McKane, R. B., Johnson, L. C., Shaver, G. R., Nadelhoffer, K. J., Rastetter, E. B., Fry, B., Giblin, A. E., Kielland, K., Kwiatkowski, B. L., Laundre, J. A., and Murray, G. 2002. Resource-based niches provide a basis for plant species diversity and dominance in arctic tundra. *Nature* 415: 68–71.

Pace, M. L., Cole, J. J., Carpenter, S. R., Kitchell, J. F., Hodgson, J. R., Van de Bogert, M. C., Bade, D. L., Krtizberg, E. L., and Bastviken, D. 2004. Whole-lake carbon-13 additions reveal terrestrial support of aquatic food webs. *Nature* 427: 240–243.

Postel, S. L., and Thompson, B. H. Jr. 2005. Watershed protection: Capturing the benefits of nature's water supply services. *Natural Resources Forum* 29: 98–108.

Schindler, D. E., Armstrong, J. B., and Reed, T. E. 2015. The portfolio concept in ecology and evolution. *Frontiers in Ecology and the Environment* 13: 257–263.

Shiklomanov, I. 1993. World fresh water resources. In *Water in Crisis: A Guide to the World's Fresh Water Resources* (P. H. Gleick, editor). Oxford University Press, New York.

Srivastava, D. S., and Vellend, M. 2005. Biodiversity-ecosystem function research: Is it relevant to conservation? *Annual Review of Ecology, Evolution and Systematics* 36: 267–294.

Tilman, D., Knops, J., Wedin, D., and Reich, P. 2001. Experimental and observational studies of diversity, productivity and stability. In *The Functional Consequences of Biodiversity: Empirical Progress and Theoretical Extensions* (A. P. Kinzig, S. W. Pacala, and D. Tilman, editors). Princeton University Press, Princeton, NJ.

Vaughn, C. C. 2010. Biodiversity losses and ecosystem function in freshwaters: Emerging conclusions and research directions. *BioScience* 60: 25–35.

CHAPTER 4

Allendorf, F. W., and Hard, J. J. 2009. Human-induced evolution cause by unnatural selection through harvest of wild animals. *Proceedings of the National Academy of Science USA* 106: 9987–9994.

Bassar, R. D., Marshall, M. C., López-Sepulcre, A., Zandona, E., Auer, S. K., Travis, J., Pringle, C. M., Flecker, A. S., Thomas, S. A., Fraser, D. F., and Reznick, D. N. 2010. Local adaptation in Trinidadian guppies alters ecosystem processes. *Proceedings of the National Academy of Science USA* 107: 3616–3621.

Darimont, C. T., Carlson, S. M., Kinnison, M. T., Paquet, P. C., Reimchen, T. E., and Wilmers, C. C. 2009. Human predators outpace other agents of trait change in the wild. *Proceedings of the National Academy of Science USA* 106: 952–954.

Foley, J. A., DeFries, R., Asner, G. P., Barford, C., Bonan, G., Carpenter, S. R., Chapin, F. S. III, Coe, M. T., Daily, G. C., Gibbs, H. K., Helkowski, J. H., Holloway, T., Howard, E. A., Kucharik, C. J., Monfreda, C., Patz, J. A., Prentice, I. C., Ramankutty, N., and Snyder, P. K. 2005. Global consequences of land use. *Science* 309: 570–573.

Golley, F. B. 1991. The ecosystem concept: A search for order. *Ecological Research* 6: 129–138.

Grant, G. C., and Scholes, M. C. 2006. The importance of nutrient hot-spots in the conservation and management of large wild mammalian herbivores in semi-arid savannas. *Biological Conservation* 130: 426–437.

Haberl, H., Erb, K. H., Krausman, F., Gaube, V., Bondeau, A., Putzar, C., Gingrich, S., Lucht, W., and Fischer-Kolwalski, M. 2007. Quantifying and mapping the human appropriation of net primary production in earth's terrestrial ecosystems. *Proceedings of the National Academy of Science USA* 104: 12942–12947.

Haddad, N. M., Brudvig, L. A., Clbert, J., Davies, K. F., Gonzalez, A., Holt, R. D., Loverjoy, T. E., Sexton, J. O., Austin, M., Collins, C. D., Cook, W. M., Damschen, E. I., Ewers, R. M., Foster, B. L., Jenkins, C. N., King, A. J., Laurance, W. F., Levey, D. J., Margules, C. R., Melbourne, B. A., Nicholls, A. O., Orrock, J. L., Song, D-X., and Townsend, J. R. 2015. Habitat fragmentation and its lasting impact on Earth's ecosystems. *Science Advances* 1: e1500052.

Hastings, A., Byers, J. E., Crooks, J. A., Cuddington, K., Jones, C. G., Lambrions, J. G., Talley, T. S., and Wilson, W. G. 2006. Ecosystem engineering in space and time. *Ecology Letters* 10: 153–164.

Holt, R. D. 1995. Linking species and ecosystems: Where is Darwin? In *Linking Species and Ecosystems.* (C. Jones and J. H. Lawton, editors). Chapman and Hall, London.

Imhoff, M. L., Bounoua, L., Ricketts, T., Loucks, C., Hariss, R., and Lawrence, W. T. 2004. Global patterns in human consumption of net primary production. *Nature* 429: 870–873.

Jones, C. G., Lawton, J. H., and Shachak, M. 1994. Organisms as ecosystem engineers. *Oikos* 69: 373–386.

Kareiva, P., and Wennergren, U. 1995. Connecting landscape patterns to ecosystem and population processes. *Nature* 373: 299–373.

Krausmann, F., Erb, K-H., Gingrich, S., Haberl, H., Bondeau, A., Gaube, V., Lauk, C., Plutzar, C., and Searchinger, T. D. 2013. Global human appropriation of net primary productions in the twentieth century. *Proceedings of the National Academy of Science USA* 110: 10324–10329.

Leibold, M. A., Holyoak, M., Mouquet, N., Amaresekere, P., Chase, J. M., Hoopes, M. F., Holt, R. D., Shurin, J. B., Lae, R., Tilman, D., Loreau, M., and Gonzalez, A. 2004. The metacommunity concept: A framework for multiscale community ecology. *Ecology Letters* 7: 601–613.

Louv, R. 2005. *Last Child in the Woods: Saving Our Children from Nature-deficit Disorder.* Algonquin Books, Chapel Hill, NC.

Martinson, H. M., and Fagan, W. F. 2014. Trophic disruption: A meta-analysis of how habitat fragmentation affects resource consumption in terrestrial systems. *Ecology Letters* 17: 1178–1189.

Naiman, R. J., Johnston, C. A., and Kelley, J. C. 1988. Alteration of North American streams by beaver. *BioScience* 38: 753–761.

Olsen, E. M., Heino, M., Lilly, G. R., Morgan, M. J., Brattey, J., Ernande, B., and Dieckman, U. 2004. Maturation trends indicative of rapid evolution preceded the collapse of the northern cod. *Nature* 428: 932–935.

O'Neill, R. V. 2001. Is it time to bury the ecosystem concept? (With full military honors, of course). *Ecology* 82: 3275–3284.

Polis, G. A., and Hurd, S. D. 1995. Extraordinarily high densities of spiders on islands: Flow of energy from the marine to terrestrial food webs and the absence of predation. *Proceedings of the National Academy of Science USA* 92: 4382–4386.

Polis, G. A., Power, M. E., and Huxel, G. R. (editors). 2004. *Food Webs at the Landscape Level.* University of Chicago Press, Chicago.

Pringle, R. M., Doak, D. F., Palmer, T. M., Jocque, R., and Brody, A. K. 2010. Spatial pattern enhances ecosystem functioning in an African savanna. *PLoS Biol* 85(5): e1000377.

Reznick, D. N., Shaw, F. H., Rodd, F. H., and Shaw, R. 1997. Evaluation of the rate of evolution in natural populations of guppies (Poecilia reticulate). *Science* 275: 1934–1937.

Schmitz, O. J., Hawlena, D., and Trussell, G. R. 2010. Predator control of ecosystem nutrient dynamics. *Ecology Letters* 13: 1199–1209.

Schmitz, O. J. 2013. Terrestrial food webs and vulnerability of the structure and functioning of ecosystems to climate. In *Climate Vulnerability: Understanding and Addressing Threats to Resources* (R. Pielke Sr., T. Seastedt, and K. Suding, editors). Academic Press, Elsevier, Cambridge, MA.

Schoener, T. W. 2011. The newest synthesis: understanding the interplay of evolutionary and ecological dynamics. *Science* 331: 426–429.

Seto, K. C., Bringezu, S., deGroot, R., Erb, K., Graedel, T. G., Ramankutty, N., Reenberg, A., Schmitz, O. J., and Skole, D. 2009. Land: Stocks, flows and prospects. In *Linkages of Sustainability* (T. Graedel and E. van der Voet, editors). Strüngman Forum Report, volume 4. MIT Press, Cambridge, MA.

Sinclair, A.R.E., Mduma, S., and Brashares, J. S. 2003. Patterns of predation in a diverse predator-prey system. *Nature* 425: 288–290.

Strong, D. R., and Frank, K. T. 2010. Human involvement in food webs. *Annual Review of Environment and Resources* 35: 1–23.

Vander Zanden, M. J., Shuter, B. J., Lester, N., and Rasmussen, J. B. 1999. Patterns of food chain length in lakes: A stable isotope study. *American Naturalist* 154: 406–416.

Vanni, M. J. 2002. Nutrient cycling by animals in freshwater ecosystems. *Annual Review of Ecology and Systematics* 33: 341–370.

Vitousek, P. M., Ehrlich, P. R., Ehrlich, A. H., and Mattson, P. A. 1986. Human appropriation of the products of photosynthesis. *BioScience* 36: 368–373.

Wilcove, D. S., Rothstein, D., Dubow, J., Phillips, A., and Losos, E. 1998. Quantifying threats to imperiled species in the United States. *BioScience* 48: 607–615.

CHAPTER 5

Beaugrand, G., Edwards, M., and Legendre, L. 2010. Marine biodiversity, ecosystem functioning, and carbon cycles. *Proceedings of the National Academy of Science USA* 107: 10120–10124.

Beddington, J. R., Agnew, D. J., and Clark, C. W. 2007. Current problems in the management of marine fisheries. *Science* 316: 1713–1716.

Brashares, J. S., Arcese, P., Sam, M. K., Coppolillo, P. B., Sincalir, A.R.E., and Balmford, A. 2004. Bushmeat hunting, wildlife declines and fish supply in West Africa. *Science* 306: 1180–1183.

Carpenter, S. R., and Brock, W. A. 2006. Rising variance: A leading indicator of ecological transition. *Ecology Letters* 9: 308–315.

Carpenter, S. R., Brock, W. A., Cole, J. J., Kitchell, J. F., and Pace, M. L. 2008. Leading indicators of trophic cascades. *Ecology Letters* 11: 128–138.

Chapin, F. S. III, Folke, C., and Kofinas, G. P. 2009. A framework for understanding change. In *Principles of Ecosystem Stewardship: Resilience-based Natural Resource Management in a Changing World* (F. S. Chapin, G. Kofinas, and C. Folke, editors). Springer, New York.

Commoner, B. 1971. *The Closing Circle: Nature, Man and Technology*. Random House, New York.

Dawson, T. P., Rounsevell, M.D.A., Kluvánková-Oravská, T., Chobotová, V., and Stirling, A. 2010. Dynamic properties of complex adaptive systems: implications for the sustainability of service provision. *Biodiversity Conservation* 19: 2843–2853.

Dobson, A. P., Lodge, D., Adler, J., Cumming, G. S., Keymer, J., McGlade, J., Mooney, H., Rusak, J. A., Sala, O., Wolters, V., Wall, D., Winfree, R., and Xenopoulos, M. A. 2009. Habitat loss, trophic collapse, and the decline of ecosystem services. *Ecology* 87: 1915–1924.

Folke, C., Carpenter, S. R., Elmqvist, T., Gunderson, L., Holling, C. S., and Walker, W. 2002. Resilience and sustainable development: building adaptive capacity in a world of transformations. *Ambio* 31: 437–440.

Frank, K. T., Petrie, B., and Shackell, N. L. 2007. The ups and downs of trophic control in continental shelf ecosystems. *Trends in Ecology and Evolution* 22: 236–242.

Gunderson, L. H. 2000. Ecological resilience—in theory and application. *Annual Review of Ecology and Systematics* 31: 425–439.

Hutchings, J. A., and Myers, R. A. 1995. The biological collapse of Atlantic cod off Newfoundland and Labrador: An exploration of historical changes in exploitation, harvesting technology, and management. In *The North Atlantic Fishery: Strengths, Weaknesses, and Challenges* (R. Arnason and L. F. Felt, editors). Institute of Island Studies, University of Prince Edward Island, Charlottetown, PEI.

Keohane, N. O., and Olmstead, S. M. 2007. Chapter 7: Stocks that grow: The economics of renewable resource management. In *Markets and the Environment*. Island Press, Washington, DC.

Kricher, J. 2009. *The Balance of Nature: Ecology's Enduring Myth*. Princeton University Press, Princeton, NJ.

Lear, W. H. 1998. History of fisheries in the northwest Atlantic. *Journal of Northwest Atlantic Fisheries Science* 23: 41–73.

Levin, S. A. 1998. Ecosystems and the biosphere as complex adaptive systems. *Ecosystems* 1: 431–436.

Levin, S. A., Xepapadeas, T., Crépin, A-S., Norberg, J., de Zeeuw, A., Folke, C., Hughes, T., Arrow, K., Barrett, S., Daily, G., Ehrlich, P., Kautsky, N., Mäler, K-G., Polasky, S., Troell, M., Vincent, J. R., and Walker, B. 2013. Social-ecological systems as complex adaptive systems: Modeling and policy implications. *Environment and Development Economics* 18: 111–132.

Liu, J., Dietz, T., Carpenter, S. R., Alberti, M., Folke, C., Morna, E., Pell, A. N., Deadman, P., Kratz, T., Lubchenco, J., Ostrom, E., Ouyang, Z., Provencher, W., Redman, C. L., Schneider, S. H., and Taylor, W. W. 2007. Complexity of coupled-human and natural systems. *Science* 317: 1513–1516.

Machlis, G. E., Force, J. E., and Burch, W. R. 1997. The human ecosystem Part I: The human ecosystem as an organizing concept in ecosystem management. *Society and Natural Resources* 10: 347–367.

Meadows, D. H. 2008. *Thinking in Systems.* Chelsea Green Publishing Company, White River Junction, VT.

O'Neill, R. V., and Kahn, J. R. 2000. Homo economous as a keystone species. *BioScience* 50: 333–337.

Rowcliffe, J. M., Milner-Gulland, E. J., and Cowlinshaw, M. 2005. Do bushmeat consumers have other fish to fry? *Trends in Ecology and Evolution* 20: 274–276.

Schmitz, O. J. 2005. Scaling from plot experiments to landscapes: Studying grasshoppers to inform forest ecosystem management. *Oecologia* 145: 225–234.

Schmitz, O. J. 2009. Perspectives on sustainability of ecosystem services and functions. In *Linkages of Sustainability* (T. Graedel and E. van der Voet, editors). Strüngman Forum Report, volume 4. MIT Press, Cambridge, MA.

Schmitz, O. J. 2010. *Resolving Ecosystem Complexity.* Princeton University Press, Princeton, NJ.

Worm, B., Hilborn, R., Baum, J. K., Branch, T. A., Collie, J. S., Costello, C., Fulton, E. A., Hutchings, J. A., Jennings, S., Jensen, O. P., Lotze, H. K., Mace, P. M., McClanahan, T. R., Minto, C., Palumbi, S. R., Parma, A. M., Ricard, D., Rosenberg, A. W., Watson, R., and Zel, D. 2009. Rebuilding global fisheries. *Science* 325: 578–584.

CHAPTER 6

Beschta. R. L., and Ripple, W. J. 2006. River channel dynamics following extirpation of wolves in northwestern Yellowstone National Park, USA. *Earth Surface Processes and Landforms* 31: 1525–1539.

Chapin, F. S. III, Kofinas, G. P., and Folke, C. (editors). 2009. *Principles of Ecosystem Stewardship: Resilience-based Natural Resource Management in a Changing World*. Springer, New York.

Chapin, F. S. III, Power, M. E., Pickett, S.T.A., Freitag, A., Reynolds, J. A., Jackson, R. B., Lodge, D. M., Duke, C., Collins, S. L., Power, A. G., and Bartuska, A. 2011. Earth stewardship: Science for action to sustain the human-earth system. *Ecosphere* 2: art89 doi 10.1890/ES11-00166.1.

Costanza, R., de Groot, R., Sutton, P., van der Ploeg, S., Anderson, S. J., Kubiszewski, I., Farber, S., and Turner, R. K. 2014. Changes in global value of ecosystem services. *Global Environmental Change* 26: 152–158.

Dawson, T. P., Rounsevell, M.D.A., Kluvánková-Oravská, T., Chobotová, V., and Stirling, A. 2010. Dynamic properties of complex adaptive systems: Implications for the sustainability of service provision. *Biodiversity Conservation* 19: 2843–2853.

Hobbs, R. J., and Harris, J. A. 2001. Restoration ecology: Repairing the Earth's ecosystems in the new millennium. *Restoration Ecology* 9: 239–246.

Howden, M. S., Soussana,J-F., Tubiello, F. N., Chhetri, N., Dunlop, M., and Meinke, H. 2007. Adapting agriculture to climate change. *Proceedings of the National Academy of Science USA* 104: 19691–19696.

Jeffries, R. J., Rockwell, R. F., and Abraham, K. F. 2004. Agricultural food subsidies, migratory connectivity and large scale disturbance in arctic coastal systems: A case study. *Integrative and Comparative Biology* 44: 130–139.

Jones, H. P., and Schmitz, O. J. 2009. Rapid recovery of damaged ecosystems. *Plos One* 4: e5653. doi:10.1371/journal.pone.0005653.

Kellert, S. R., and Wilson, E. O. 1993. *The Biophilia Hypothesis*. Island Press, Washington, DC.

Levin, S. A., Xepapadeas, T., Crépin, A-S., Norberg, J., de Zeeuw, A., Folke, C., Hughes, T., Arrow, K., Barrett, S.,

Daily, G., Ehrlich, P., Kautsky, N., Mäler, K-G., Polasky, S., Troell, M., Vincent, J. R., and Walker, B. 2013. Social-ecological systems as complex adaptive systems: Modeling and policy implications. *Environment and Development Economics* 18: 111–132.

Liu, J., Mooney, H., Hull, V., Davis, S. J., Gaskell, J., Hertel T., Lubchenco, J., Seto, K. C., Glieck, P., Kremen, C., and Li, S. 2015. Systems integration for global sustainability. *Science* 347, 1258832. doi 10.1126/science.1258832.

Mace, G. M. 2014. Whose conservation? *Science* 345: 1558–1560.

Marshall, K. N., Hobbs, N. T., and Cooper, D. J. 2013. Stream hydrology limits recovery of riparian ecosystems after wolf reintroduction. *Proceedings of the Royal Society* B 280: 20122977.

Palmer, M. A., Bernhardt, E. S., Schlesinger, W. H., Eshleman, K. N., Foufoula-Georgiou, E., Hendryx, M. S., Lemly, A. D., Likens, G. E., Loucks, O. L., Power, M. E., White, P. S., and Wilcock, P. S. 2010. Mountaintop mining consequences. *Science* 327: 148–149.

Palmer, M. A., and Filoso, S. 2009. Restoration of ecosystem services for environmental markets. *Science* 325: 575–576.

Rey Benayas, J. M., Newton, A. C., Diaz, A., and Bullock, J. M. 2009. Enhancement of biodiversity and ecosystem services by ecological restoration: A meta-analysis. *Science* 325: 1121–1124.

Rockstrom, J., Steffen, W. Noone, K., Persson, A., Chapin, F. S. III, Lambin, E. F., Lenton, T. M., Scheffer, M., Folke, C., Schllnuber, H. J., Nykvist, B., de Witt, C. A., Hughes, T., van der Leeuw, S., Rodhe, H., Sörlin, S., Snyder, P. K., Costanza, R., Svedin, U., Falkenmark, M., Karlberg, L., Corell, R. W., Fabry, V. J., Hansen, J., Walker, B., Liverman, D., Richardson, K., Crtutzen, P., and Floey, J. A. 2009. A safe operating space for humanity. *Nature* 461: 472–475.

Schmitz, O. J. 2009. Perspectives on sustainability of ecosystem services and functions. In *Linkages of Sustainability* (T. Graedel and E. van der Voet, editors). Strüngman Forum Report, volume 4. MIT Press, Cambridge, MA.

Schmitz, O. J. 2013. Terrestrial food webs and vulnerability of the structure and functioning of ecosystems to climate. In *Climate Vulnerability: Understanding and Addressing Threats to Resources* (R. Pielke Sr., T. Seastedt, and K. Suding, editors). Academic Press, Elsevier, Cambridge, MA.

Suding, K. N., and Hobbs, R. J. 2009. Threshold models in restoration and conservation: A developing framework. *Trends in Ecology and Evolution* 24: 271–279.

Wilson, E. O. 1984. *Biophilia*. Harvard University Press, Cambridge, MA.

CHAPTER 7

Allenby, B. 1998. Earth systems engineering: The role of industrial ecology in an engineered world. *Journal of Industrial Ecology* 2: 73–93.

Chapin, F. S. III, Folke, C., and Kofinas, G. P. 2009. A framework for understanding change. In *Principles of Ecosystem Stewardship: Resilience-based Natural Resource Management in a Changing World* (F. S. Chapin, G. Kofinas, and C. Folke, editors). Springer, New York.

Chen, W.-Q., and Graedel, T. E. 2012. Anthropogenic cycles of the elements: A critical review. *Environmental Science and Technology* 46: 8574–8586.

Committee on Critical Mineral Impacts on the U.S. Economy. 2008. *Minerals, Critical Minerals, and the U.S. Economy.* National Academy Press, Washington, DC.

DeAngelis, D. L., Mullholland, P. J., Palumbo, A. V., Steinman, A. D., Huston, M. A., and Ellwood, J. W. 1989.

Nutrient dynamics and food web stability. *Annual Review of Ecology and Systematics* 20: 71–95.

Dodds, W. K. 2008. *Humanity's Footprint: Momentum, Impact, and Our Global Environment*. Columbia University Press, New York.

Felson, A. J., and Pickett, S.T.A. 2005. Designed experiments: New approaches to studying urban ecosystems. *Frontiers in Ecology and the Environment* 3: 549–556.

Gómez-Baggethun, E., and Barton, D. N. 2013. Classifying and valuing ecosystem services for urban planning. *Ecological Economics* 86: 235–245.

Gordon, R. B., Bertram, M., and Graedel, T. E. 2006. Metal stocks and sustainability. *Proceedings of the National Academy of Science USA* 102: 1209–1214.

Graedel, T. E. 1996. On the concept of industrial ecology. *Annual Review of Energy and the Environment* 21: 69–98.

Grove, J. M. 2009. Cities: Managing densely settled socioecological systems. In *Principles of Ecosystem Stewardship: Resilience-based Natural Resource Management in a Changing World* (F. S. Chapin, G. Kofinas, and C. Folke, editors). Springer, New York.

Grove, J. M., Cadenasso, M., Pickett, S.T.A., Machlis, G. E., and Burch, W. R. 2015. *The Baltimore School of Urban Ecology: Space, Scale and the Time for the Study of Cities*. Yale University Press, New Haven, CT.

Holling, C. S. 2002. *Panarchy: Understanding Transformations in Human and Natural Systems*. Island Press, Washington DC.

Hoornweg, D., Bhada-Tata, P., and Kennedy, C. 2013. Environment: Waste production must peak this century. *Nature* 502: 615–617.

Kardan, O., Gozdyra, P., Misic, B., Moola, F., Plamer, L. J., Paus, T., and Berman, M. G. 2015. Neighborhood greenspace and health in a large urban center. *Scientific Reports* 5: 11610.

Kean, S. 2010. Fishing for gold in the last frontier state. *Science* 327: 263–265.

Levin, S. A., Barrett, S., Aniyar, S., Baumol, W., Bliss, C., Bolin, B., Dasgupta, P., Ehrlich, P., Folke, C., Gren, I.-M., Holling, C. S., Jansson, A., Jansson, B.-O., Martin, D., Maler, K.-G., Perrings, C., and Sheshinsky, E. 1998. Resilience in natural and socioeconomic systems. *Environmental and Developmental Economics* 3: 221–262.

Liu, J., Mooney, H., Hull, V., Davis, S. J., Gaskell, J., Hertel, T., Lubchenco, J., Seto, K. C., Glieck, P., Kremen, C., and Li, S. 2015. Systems integration for global sustainability. *Science* 347, 1258832. doi 10.1126/science.1258832.

Lombardi, R. D., and Laybourn, P. 2012. Redefining industrial symbiosis. *Journal of Industrial Ecology* 16: 28–37.

Loreau, M. 1995. Consumers as maximizers of matter and energy flow in ecosystems. *American Naturalist* 145: 22–42.

Pickett, S.T.A., and Cadenasso, M. 2008. Linking ecological and built components of urban mosaics: An open cycle of ecological design. *Journal of Ecology* 96: 8–12.

Schmitz, O. J. 2009. Perspectives on sustainability of ecosystem services and functions. In *Linkages of Sustainability* (T. Graedel and E. van der Voet, editors). Strüngman Forum Report, volume 4. MIT Press, Cambridge, MA.

Schmitz, O. J., and Graedel, T. E. 2010. The consumption conundrum: Driving destruction abroad. Yale E360 http://e360.yale.edu/content/feature.msp?id=2266.

Schumpeter, J. A. 2008. *Capitalism, Socialism and Democracy*, 3rd edition. Harper Perennial Modern Classics, New York.

Seitzinger, S., Svedin, U., Crumley, C. L., Steffen, W., Abdullah, S. A., Alfsen, C., Broadgate, W. J., Biermann, F., Bondre, N. R., Dearing, J. A., Deutsch, L., Dhakal, S., Elmqvist, T., Farahbakhshazad, N., Gaffney, O., Haberl, H., Lavorel, S., Mbow, C., McMichael, A. J., deMorais, J.M.F., Olsson, P., Pinho, P. F., Seto, K. C., Sinclair, P., Stafford Smith, M. S., and Sugar, L. 2012. Planetary stewardship in an urbanizing world: Beyond city limits. *Ambio* 41: 787–794.

Tanner, C. J., Adler, F. R., Grimm, N. B., Groffman, P. M., Levin, S. A., Munshi-South, J., Pataki, D. E., Pavo-Zuckerman, M., and Wilson, W. G. 2014. Urban ecology: Advancing science and society. *Frontiers in Ecology and Environment* 12: 574–581.

Wu, J. 2014. Urban ecology and sustainability: The state-of-the-science and future directions. *Landscape and Urban Planning* 125: 209–221.

Zheng, H., Robinson, B. E., Lian, Y.-C., Polasky, S., Ma, D.-C., Wang, F.-C., Ruckelshaus, M., Ouyang, Z.-Y., and Daily, G. C. 2013. Benefits, costs and regional implications of a regional payment for ecosystems services program. *Proceedings of the National Academy of Science USA* 110: 16681–16686.

CHAPTER 8

Alberti, M. 2015. Eco-evolutionary dynamics in an urbanizing planet. *Trends in Ecology and Evolution* 30: 114–126.

Donihue, C. M., and Lambert, M. R. 2015. Adaptive evolution in urban systems. *Ambio* 44: 194–203.

Jetz, W., McPeherson, J. M., and Guralinick, R. P. 2012. Integrating biodiversity distribution knowledge: Toward a global map of life. *Trends in Ecology and Evolution* 27: 151–159.

Leopold, A. 1966. Song of the Gavilan (page 153) in *A Sand County Almanac with Essays on Conservation from Round River*. Oxford University Press, Oxford, UK.

Levin, S. A. 1999. *Fragile Dominion: Complexity and the Commons*. Perseus Publishing, Cambridge, MA.

Mace, G. M. 2014. Whose conservation? *Science* 345:1558–1560.

McGill, B. J., Dornela, M., Gotelli, N. J., and Magurran, A. E. 2015. Fifteen forms of biodiversity trend in the Anthropocene. *Trends in Ecology and Evolution* 30: 104–113.
Naeem, S., Duffy, J. E., and Zavaletta, E. 2012. The functions of biological diversity in an age of extinction. *Science* 336: 1401–1406.
Schmitz, O. J. 2007. *Ecology and Ecosystem Conservation.* Island Press, Washington DC.

INDEX